Adam Smith's Political Pl

When Adam Smith published his celebrated writings on economics and moral philosophy he famously referred to the operation of an invisible hand. *Adam Smith's Political Philosophy* makes visible the invisible hand by examining its significance in Smith's political philosophy and relating it to similar concepts used by other philosophers, revealing a distinctive approach to social theory that stresses the significance of the unintended consequences of human action.

This book introduces greater conceptual clarity to the discussion of the invisible hand and the related concept of unintended order in the work of Smith and in political theory more generally. By examining the application of spontaneous order ideas in the work of Smith, Hume, Hayek and Popper, *Adam Smith's Political Philosophy* traces similarities in approach and from these builds a conceptual, composite model of an invisible hand argument. While setting out a clear model of the idea of spontaneous order the book also builds the case for using the idea of spontaneous order as an explanatory social theory, with chapters on its application in the fields of science, moral philosophy, law and government.

Craig Smith is a British Academy Postdoctoral Fellow in the Department of Politics at the University of Glasgow where he is conducting research on the political thought of Adam Ferguson.

Routledge studies in social and political thought

Adam Smith's Political Philosophy

The invisible hand and spontaneous order

Craig Smith

Routledge
Taylor & Francis Group

LONDON AND NEW YORK

First published 2006
by Routledge

2 Park Square, Milton Park, Abingdon, Oxon OX14 4RN (UK)

Simultaneously published in the USA and Canada
by Routledge

711 Third Avenue, New York, NY 10017 (US)

First issued in paperback 2013

Routledge is an imprint of the Taylor & Francis Group, an informa business

© 2006 Craig Smith

Typeset in Garamond by Wearset Ltd, Boldon, Tyne and Wear

British Library Cataloguing in Publication Data
A catalogue record for this book is available from the British Library

Library of Congress Cataloging in Publication Data
A catalog record for this book has been requested

ISBN 0-415-36094-3
ISBN 978-0-415-84584-7 (Paperback)

Contents

Preface

This book began life as part of the research for my PhD thesis entitled *The Idea of Spontaneous Order in Liberal Political Thought*. My research has benefited throughout from the patient guidance and advice of Professor Christopher J. Berry and from the supportive atmosphere of the Department of Politics at the University of Glasgow. Thanks are also due to my examiners, Professor Norman Barry and Mr Michael Lessnoff for many helpful comments. The research was undertaken with the support of the Student's Awards Agency for Scotland's Scottish Studentship Scheme and the editorial work was completed during my appointment as a British Academy Postdoctoral Fellow. I'm grateful to both of these institutions for their support, and also to the Institute for Humane Studies, the Institute of Economic Affairs and the Liberty Fund for the opportunity to discuss topics relating to the research. Terry Clague and the editorial staff at Routledge Research are also to be thanked for their work in producing this volume and I'm grateful to the Oxford University Press for permission to reproduce material from the Glasgow Edition of the Works and Correspondence of Adam Smith. I should also mention that the material in this book is complemented by my article 'Adam Smith on Progress and Knowledge' that will appear in the Routledge volume *New Voices on Adam Smith* edited by Eric Schliesser and Leonidas Montes. Finally, on a personal note, thanks are due to Wallace and Mary Jackson and to my family, without whose support this book wouldn't have appeared, and to whom it is dedicated.

Abbreviations

ECS   Adam Ferguson *An Essay on the History of Civil Society* [1767], ed. Fania Oz-Salzberger, Cambridge: Cambridge University Press, 1995.

EMPL   David Hume *Essays Moral, Political, and Literary* [1777], ed. Eugene F. Miller, Indianapolis: Liberty Fund, 1985.

ENQ   David Hume *Enquiries Concerning Human Understanding and Concerning the Principles of Morals* [1777], 3rd edition, ed. L.A. Selby-Bigge, rev. P.H. Nidditch, Oxford: Clarendon Press, 1975.

EPS   Adam Smith *Essays on Philosophical Subjects* [1795], ed. W.P.D. Wightman, Oxford: Oxford University Press, 1980.

LJP   Adam Smith *Lectures on Jurisprudence*, ed. R.L. Meek, D.D. Raphael and P.G. Stein, Oxford: Oxford University Press, 1978.

LLL   F.A. Hayek *Law, Legislation and Liberty* [1973–82], 3 vols. London: Routledge, 1993.

LRBL   Adam Smith *Lectures on Rhetoric and Belles Lettres*, ed. J.C. Bryce, Oxford: Oxford University Press.

THN   David Hume *A Treatise of Human Nature* [1739], 2nd edition, ed. L.A. Selby-Bigge, rev. P.H. Nidditch, Oxford: Clarendon, 1978.

TMS   Adam Smith *The Theory of Moral Sentiments* [1759], ed. D.D. Raphael and A.L. Macfie, Oxford: Oxford University Press, 1976. Reproduced by permission of Oxford University Press.

WN   Adam Smith *An Inquiry into the Nature and Causes of the Wealth of Nations* [1776], ed. R.H. Campbell, A.S. Skinner and W.B. Todd, Oxford: Oxford University Press, 1976. Reproduced by permission of Oxford University Press.

1 Spontaneous order in liberal political thought

> All nature is connected; and the world itself consists of parts, which, like the stones of an arch, mutually support and are supported. This order of things consists of movements, which, in a state of counteraction and apparent disturbance, mutually regulate and balance one another.
>
> (Ferguson 1973 vol. 1: 18)

The term 'invisible hand' is perhaps the most famous phrase to have emerged from the political philosophy of Adam Smith. The significance of the concept has been the subject of much scholarly discussion and its supposed implications the target of intense critical attacks. The purpose of the present study is to attempt to make visible the invisible hand and, hopefully, to illuminate the core concept of Smith's political thought. The aim of this book is to clarify with some precision the meaning of the term invisible hand and the related, modern concept of spontaneous order. It will be argued that spontaneous order thought represents a distinctive approach to social theory; and the aim of the study will be to identify its core principles and to develop a conceptual model of this approach. By identifying the key features of a spontaneous order approach as they appear in the work of the two most significant groups of spontaneous order theorists – the Scottish Enlightenment and the twentieth-century classical liberal revival – the book will build a composite model of the application of the approach to the explanation of science, morality, law and government and the market. The analysis will concentrate on spontaneous order as a descriptive approach to social theory rather than as an offshoot of attempts to justify liberal principles. As a result it will be demonstrated that the use of spontaneous order as an explanatory social theory is prior to, and a prerequisite for, the use of invisible hand arguments to justify liberal institutions.

The notion of spontaneous order has appeared at various times down the centuries and has been applied in a variety of academic disciplines: spontaneous order-inspired arguments can be found in the fields of biology, science, epistemology, language, economics, history, law, theology, sociology, anthropology and even recently in management studies and computing.

However, this study will focus on its appearance in what may be broadly referred to as social and political theory. This field, though it to a certain extent embraces elements of many of the above, is nonetheless more focused on the application of a spontaneous order approach to social and political interaction. Though economics and economists loom large in our study, and in most discussions of the notion of spontaneous order, the aim is to concentrate on what they have to say about the political theory of spontaneous orders. That is, we will consider the market, often taken to be the paradigmatic example of a spontaneous order, as one social phenomenon among others and not purely as an economic model. For this reason our analysis will begin by examining the application of the approach in the field of science.

Our subject matter is spontaneous order in liberal political thought and, before we commence, it is necessary to make clear exactly where within the broad church of liberalism these ideas appear. The first distinction we might usefully make is between the use of the term 'liberal' as it is traditionally understood in the history of political thought, and its use in the United States as a description of a particular political position. A liberal, in this American sense, is what in Europe might be called a social democrat; liberalism in America has become a term that refers, particularly, to the left-leaning intelligentsia within the Democratic Party. We are not then talking about liberalism in this sense. Hayek, in his *Why I am not a Conservative*, argues that 'liberal', as a descriptive term, is no longer accurate as a result of this development. It does not refer to the same set of ideas as once it did, and the popularity of this new meaning in the United States makes its use misleading. What instead we are concerned with is what has come to be known as classical liberalism.

A second distinction should be made at this point: that is between Anglo-American and Continental liberalism. This distinction broadly follows that between the philosophical traditions which Popper identifies as British Empiricist and Continental Rationalist (Popper 1989: 4). The thinkers of the spontaneous order tradition take great pains to emphasize this distinction (Hayek 1979: 360; 1960: 55–7). They argue that the Cartesian-influenced constructivist rationalism of the Continental school's methodology sharply distinguishes it from the Anglo-American tradition of empirical, analytical liberalism. Spontaneous order theorists identify themselves with the Anglo-American philosophical approach to liberalism, and expend considerable energy in a critique of continental rationalist thought.

Spontaneous order theories occur within Anglo-American classical liberal thought.[1] There is, however, a further distinction which might be drawn to specify the position of spontaneous order thought within liberalism: that is a distinction between what Gissurarson (1987: 155–6), following Buchanan (1977: 38), typifies as American libertarianism and European classical liberalism. The distinguishing feature here is the rights-based contractarian approach of libertarianism in contrast to the evolutionary gradualism of classical liberalism. The contrast arises from the evolved nature of European liberalism, as opposed to the intentional constitution building of American Libertarianism. This distinction leads Gissurarson to place spontaneous order thought within a tradition that he refers to as 'conservative liberalism' (Gissurarson 1987: 6).[2] However, given the distinctions which we have drawn thus far it would be more accurate to refer to spontaneous order thought as existing in a subset of liberalism which we might call British Whig Evolutionary Liberalism, a subset whose distinguishing characteristic, as we will see, is precisely its concern with the notion of the spontaneous formation of order.

Though we will not be undertaking a historical study, our aim being to clarify a 'model' of the spontaneous order approach, rather than to trace its historical development, it is necessary nonetheless to sketch briefly the history of the tradition in order that we might select the building blocks from which our model will rise. As we pass through the list chronologically it would appear best, for the sake of accuracy, to restrict our attention to those thinkers who express a significant spontaneous order theory in our chosen field of social and political theory. By limiting our attention in such a way we will be more able successfully to draw out the essential elements in a spontaneous order argument. With this in mind we may exclude from our study some of those to whom spontaneous order ideas have been attributed.

In his article *The Tradition of Spontaneous Order* Norman Barry conducts a study of thinkers whom, he believes, have utilized spontaneous order arguments through the centuries. This he claims, following on and building upon Hayek's views, represents the tradition of spontaneous order thinking. But if Barry's group of thinkers represent a tradition, then it is a tradition in a peculiar sense of the term. That is to say a tradition is more usually considered as something that directly relates its members; something passed down from one exponent to another. This being the case the early members of the tradition to whom Hayek and Barry refer (the Spanish Schoolmen, Molina and Hale) cannot really be considered as representing members of a tradition.[3] If we follow Quentin Skinner's criteria for attributing influence – '(1) that there should be a "genuine similarity between the doctrines" of the writers; (2) that the influenced writer could only have got the relevant doctrines from his alleged creditor; (3) that there should be a low probability of the similarities being coincidental' (Condren 1985: 133, citing Skinner 1969: 26) – we will see that, though similar ideas may recur in the work of each of these individuals or groups, it would be difficult, nay impossible, to trace with any accuracy the influences of these early writers who are credited with applying the spontaneous order approach upon each other. As Hayek's sketchy contentions in his article *Dr Bernard Mandeville* show we simply lack the evidence to assert that the Spanish Schoolmen influenced Hale, who in turn influenced Mandeville. We have no real historical record, despite Hayek's best attempts, of such a connection except the recurrence of broadly similar or conceptually similar notions, and that, for the purposes of our study, does not constitute evidence enough to refer to them as members of a tradition. It is more accurate to refer to those early thinkers to whom Hayek and Barry attribute spontaneous order ideas as precursors of the tradition of spontaneous order. One other such figure whom we might consider in this light is Giambattista Vico.

Duncan Forbes (1954: 658–9) and others have highlighted the conceptual similarities which may link Vico (1668–1744) to the tradition of spontaneous order. Forbes cites the evidence of Vico's *Scienza Nuova Seconda*, and in particular refers to one passage as the 'locus classicus' of his concept of the 'Law of the Heterogeneity of Ends' (Forbes 1954: 658). The passage in question has clear similarities with the Scots' use of what has come to be known as unintended consequences. It reads:

> The world of nations is in fact a human creation ... Yet without a doubt this world was created by the mind of providence, which is often different, sometimes contrary, and always superior to the particular goals which people have set for themselves. Instead to preserve the human race on the earth, providence uses people's limited goals as a means of attaining greater ones.

> (Vico 1999: 489–90)

Forbes though rejects any direct relationship of influence by Vico upon the Scottish Enlightenment on the grounds of a lack of historical evidence (Forbes 1954: 658–9).[4] Indeed Vico's concept of unintended consequences, though it bears conceptual similarities to the Scots', is separated from them by his constant appeal to divine providence. For Vico his *Scienza Nuova* represents a 'demonstration of what providence has wrought in history' (Vaughan 1972: 39), it establishes divine providence as 'historical fact' (Vico 1999: 127). Though his analysis of the growth of political institutions is undertaken through an unintended consequences approach, it is also carried out under a strong conception of divine intervention through providence.

The key difference between the Scots' conception of unintended consequences and that deployed by Vico is precisely over this point. What Vico attributes to God's divine providence is precisely that which the Scots seek to explain in secular sociological terms. If indeed models of divine providence and arguments from design are to be held to have influenced the Scots, whether they accepted fully the role of a providential God or whether they simply borrowed the model and then applied it to a secular social mechanism, we cannot, with any accuracy, trace this to Vico's writings. If religious models influenced the Scots it is far more likely that their conceptions of providence would be shaped by the historical context of Presbyterian Scotland, or by the broader Enlightenment Deism of Europe. As Burke wrote, Vico 'is on the frontier between the theological and the secular interpretation of history' (Burke 1985: 61). The Scots, on the other hand, stand firmly on the secular side of this great transition and our tradition, if we are to seek historical accuracy, ought to begin on that side of the divide. In other words this leads us to Bernard Mandeville (1670–1733).

With Mandeville we are able to start the tradition of spontaneous order at a point where we have some record of influence, or at least acknowledgement of influence, and where there are more definite grounds for using the terms influence and tradition. We are on safer ground if we follow a tradition of thought which begins with Mandeville whose work was clearly an influence on the thought of the Scottish Enlightenment. Aside from the fact that several of the Scots cite him in their work, and attempt critiques of his views, we also have the evidence that one of Mandeville's chief critics, Francis Hutcheson (1694–1746), was a professor at Glasgow and a teacher of Smith.[5]

It is possible to trace a distinct connection travelling from Mandeville and Hutcheson to Smith and his friends Hume and Ferguson, which we may then extend to Smith and Ferguson's respective pupils Millar and Dugald Stewart. There is little doubt that the spontaneous order approach plays an important role in much of the thought of the Scottish Enlightenment.[6] This, however, is not to claim that the movement, if indeed it was such, held a coherent position as regards the spontaneous order approach. The bulk of this study will focus on the relationship of the thought of the major Scottish exponents of spontaneous order to the more recent thinkers of the twentieth-century classical liberal revival. Our study will concentrate on three of the Scots: David Hume (1711–76), Adam Smith (1723–90) and Adam Ferguson (1723–1816).

In addition to these figures there are also a number of 'second rank', or second generation, Scots thinkers who deploy spontaneous order approaches in their thought. Ronald Hamowy (1987) has argued that spontaneous order ideas can be traced in the work of most of the Scots thinkers of this time: this includes lesser figures such as John Millar (1735–1801), Dugald Stewart (1753–1828), Lord Kames (1696–1782) and Gilbert Stuart (1743–86). In addition spontaneous order ideas are apparent in the work of Thomas Reid (1710–96), who is traditionally thought to sit somewhat outside the mainstream of the Scottish Enlightenment, and whose use of the notion shows how widespread its influence was at the time.

Following on from the Scots we can also trace ideas of spontaneous order in the thought of Edmund Burke (1729–97). Burke was himself a leading Whig politician and was known to Hume and Smith. He also served as Rector of Glasgow University and is known to have been intimately familiar with the thought of the Scots writers. From the Scottish Enlightenment we are able to trace our connection down into the next generation of political theorists by three paths.[7]

First, we have a direct link to the late eighteenth- and early nineteenth-century economists now referred to as the classical economists, particularly Say (1767–1830) and Ricardo (1772–1823) who both developed aspects of Smith's economic analysis into a highly sophisticated abstract discipline of economic science. They pick up some of the ideas of spontaneous order but, as their focus is on economics rather than the broader field of social and political theory, this will allow us to note them and pass on. The second, and related, path of development is that which stems from Dugald Stewart to his pupil James Mill (1773–1836), also considered to be a member of the school of classical economists. From here we have a direct link to his son J.S. Mill (1806–73). Hayek has questioned the Mills' relationship to the tradition of spontaneous order because of their relation to Benthamite utilitarianism. He has argued that the younger Mill is more properly considered as an exponent of the continental style rational liberalism which we contrasted with the tradition of liberalism which produced the spontaneous order approach.[8] There are nonetheless significant spontaneous order aspects which may be detected in the younger Mill's defence of liberty, particularly in *On Liberty*.

The third path of development that leads from the Scots is that which is to be found in the nineteenth-century evolutionists. Charles Darwin's (1809–82) theory of evolution is, according to Hayek, an adaptation of the Scots' spontaneous order theories applied to biology. Hayek believed that Darwin picked up these ideas through the medium of the Scots geologist James Hutton (1726–97), a member of the broader Scottish Enlightenment, and through the influence of Hume upon his grandfather Erasmus Darwin, and then applied the approach to nature.[9] Darwin himself was not a social and political theorist and so his work is outside the scope of this study except in one feature, namely, the use by Hayek and Popper of a notion of evolution which they relate to spontaneous order and which draws on the process of natural selection formulated by Darwin. From Darwin we are able to trace a development of the spontaneous order approach through the writings of Herbert Spencer (1817–62) and T.H. Huxley (1825–95). These two thinkers are often portrayed as the leading exponents of the application of Darwinian evolution to social and political matters. Hayek, however, argues that, though Spencer in particular draws on spontaneous order ideas of evolution, he sees them through the lens of Darwinian biology. That is to say that Spencer's use of evolutionary ideas owes most of its force to its reliance on eugenics. Hayek has argued (LLL vol. 1: 23–4, 152 n. 33) that this was a false path in the development of spontaneous order ideas. His distaste for this development of spontaneous order through eugenics leads him to omit any detailed discussion of either Spencer or Huxley from his work. Hayek's rejection of this development of the spontaneous order approach is grounded on the assertion that it does not follow on accurately from the work of the Scots. Indeed part of Hayek's project is to resurrect the Scots' understanding of spontaneous order in the face of the errors of nineteenth-century evolutionists.

Moving closer to our own times spontaneous order appears in the work of the Austrian School of economists, including Menger (1840–1921), Böhm-Bawerk (1851–1914), Weiser (1851–1926) and von Mises (1881–1973). The Austrians developed a subjectivist theory of value and applied it to economics. Their chief concern was with questions of epistemology as applied to economics. The Austrians were primarily concerned with technical economics and thus fall outside the central concerns of the present study. However, it is important to note the influence of Menger's methodological thought on the social and political thought of Hayek.[10] Menger set himself the question: 'How can it be that institutions which serve the common welfare and are extremely significant for its development come into being without a common will directed toward establishing them?' (Menger 1996: 124). This concern was to shape Hayek's work in Austrian economics and his move into social and political theory.

In the second half of the twentieth century there was a renewed interest in spontaneous order ideas which saw a rejuvenation of the tradition.[11] The members of this twentieth-century revival include: Michael Polanyi (1891–1976), whose place in the tradition of spontaneous order is assured by his apparent coining of the term; Karl Popper (1902–94) and F.A. Hayek (1899–1992), perhaps the greatest twentieth-century exponent of the tradition in the social and political sphere and the one who here will receive the most treatment. In addition to this triumvirate spontaneous order ideas can also be traced in the work of the conservative political theorist Michael Oakeshott (1901–90). Oakeshott's ideas are often compared to those of his near contemporary Hayek, and his theory of civil association and his views on the market possess significant enough similarities to those of Hayek to warrant his inclusion.

In his *Anarchy, State and Utopia* Robert Nozick (1938–2002) makes use of spontaneous order arguments under the name of invisible-hand arguments. A discussion of Nozick's use of this term will form part of the next section but we can note here that, although Nozick does make use of the spontaneous order approach, the bulk of his theory is rather conducted in terms of Lockean rights theory. As a result, he is perhaps more comfortably at home in the American libertarian tradition which we have distinguished from the more spontaneous order orientated British classical liberal tradition. More recently Virginia Postrel, in her *The Future and its Enemies* (1998), defends classical liberal values in a manner which is informed by spontaneous order methodology. Drawing on concrete examples from contemporary life Postrel attempts to illustrate the benefits of an open-ended future where progress is achieved through the medium of spontaneous development and freedom.

We are now in a position to identify the particular role played by the 'tradition' in this study. From our survey we are able to identify two main time periods in which the spontaneous order approach has been deployed in a significant manner to social and political issues. We might refer to them in shorthand as the Scots and the Moderns. The Scots being those major Scottish Enlightenment thinkers who deploy the spontaneous order approach: Hume, Smith, Ferguson and Millar (and by association Mandeville and Hutcheson). And the Moderns being those theorists of the twentieth-century classical liberal revival: including Hayek, Popper and Polanyi (and by association Oakeshott). The rest of this study will be devoted to an examination of the use of the concept of spontaneous order by the Scots and the Moderns; and will consist of an attempt to define coherently the nature of spontaneous order and its application as an approach to the study of social phenomena. With this aim in view the study will not concern itself with strict contextual readings of either period.

This is not to say that such an approach would be without merit, indeed contextual readings are the dominant technique in the contemporary discipline of the history of political thought; rather, it is to say that we have a different end in view. If we are to construct a descriptive 'model' of the spontaneous order approach to social and political theory, then context must necessarily take something of a back seat. Equally this is not to say that we are advancing an ahistorical abstraction. That is not our purpose: we are using the two groups of thinkers to develop a conceptual model in order to clarify a distinct approach to social theory. In this way it does not really matter if a contextual study reveals that Smith did not think what Hayek thought he did, for the evolution of the idea depends on what Hayek thought Smith thought, and not what he actually did think. Our purpose is not to criticize the modern thinkers' reading of the Scots, nor is it to highlight inconsistencies between the groups or within the groups.[12] If we are to develop a clear understanding of what the spontaneous order approach looks like, then we must seek the theoretical similarities and from there develop our analysis of the model that results. What the analysis depends upon is not so much the strictures of a tradition of direct influence, though we have made the case for such in selecting our two groups, nor does it depend upon a claim as to the accuracy of one thinker's reading of the work of a predecessor: rather it is concerned with the 'family resemblances' (Gissurarson 1987: 10) which will allow us to draw out the implications of a spontaneous order approach.[13] Oakeshott argues that this is precisely the true nature of a tradition:

> These I call tradition because it belongs to the nature of a tradition to tolerate and unite an internal variety, not insisting upon conformity to a single character, and, because, further, it has the ability to change without losing its identity.
>
> (Oakeshott 1991: 227)

With these strictures in mind the main body of the text will represent an examination of the two groups of thinkers' approaches to the same key features of social life. In the case of each group a particular thinker, Smith for the Scots and Hayek for the Moderns, will form the spine of our model. Moreover, what is stressed in this study is spontaneous order as an approach: spontaneous order as 'a methodological tool rather than an ethical postulate' (Gissurarson 1987: 42), or as a 'value-free explanatory system' (Gray 1986: 119–20) and as being 'technical rather than value based' (Shearmur 1996a: 4). Kley, referring to Hayek, notes that he does not provide a political philosophy but rather a 'distinct body of descriptive and explanatory theory' (Kley 1994: 3) which may, or may not, lead to the normative conclusions which he wishes to draw from it.[14] Our aim is to identify the spontaneous order approach to social science with a measure of clarity that might allow the examination of its relation to instrumentalist justifications of normative positions regarding freedom and liberalism.[15] To make this distinction clear we will approach the subject by way of a distinct critical vocabulary. There are certain key descriptive terms which have come to be applied to the species of argument which we will be examining. Four such terms can be identified from the literature surrounding the tradition.

Unintended consequences

The term unintended consequences is applied, most often, to the ideas expressed by the writers of the Scottish Enlightenment. It is conceptually related to the term spontaneous order in that spontaneous orders are, by Hayek and Polanyi's definition, brought about in the social sphere by a process of unintended consequences. Indeed both Hayek and Popper refer to the notion of unintended consequences as being the central subject matter of all social sciences: a claim that we shall examine later.

The key idea here refers to purpose and intentionality. An order which is created spontaneously is not the realization of an actor's intention (or purpose), rather it is the result of a process which sees the interaction of various actors pursuing different purposes. In a spontaneous order nobody can be considered to have intended the resulting order, it is, to paraphrase Ferguson, 'The result of human action, but not the result of any human design.' The term unintended consequences refers to the notion that actions create consequences other than those which are explicitly intended. This is how most social scientists, in particular Robert Merton (1976), have dealt with the issue.[16] All actions produce unintended results in the social sphere because they necessarily entail interactions and reactions that cannot fully be predicted.[17]

Duncan Forbes, in his *Scientific Whiggism: Adam Smith and John Millar*, deploys a concept of the law of the heterogeneity of ends – from Wündt's 'das Gesetz der Heterogonie der Zwecke' (Forbes 1954: 651) – to describe the notion of unintended consequences. The law of the heterogeneity of ends is described in terms of an opposition to the great man theory of history, and indeed to all historical approaches which rely on rationalistic analysis of history in terms of conscious action. By viewing the law of the heterogeneity of ends as a historical methodology, Forbes does indeed leave open the possibility of good and bad unintended consequences, just as he does not restrict the law's field of application to any particular aspect of human action. Similarly, Boudon draws distinct subsets of unintended consequence arguments, one of which he describes as that which 'may produce collective advantages that had not been explicitly sought (the "invisible hand" of Adam Smith)'

(Boudon 1982: 6). Unintended consequences arguments of this sort are, he believes, a species of dialectical argument which, through Smith's invisible hand and Hegel's cunning of reason, influenced Marx's concept of the dialectic as well as Hayek's view of rationality (Boudon 1982: 167).[18] Hamowy, however, makes a pertinent point in this connection. He argues that both Merton and Forbes 'appear to include the whole spectrum of unintended outcomes within the concept' whereas the theory of spontaneous order 'refers only to those acts the unanticipated results of which issue in complex social patterns' (Hamowy 1987: 4). Thus the term unintended consequences as it appears within the tradition of thought with which we are concerned refers not to the broad assertion that actions can produce unanticipated results, but to a more specific notion that complex orders are formed without the purposive intention of any one actor.

Having accepted this technical understanding of the term we are left, however, with a further qualification. As Hayek (LLL vol. 1: 59) himself admits, and more forcefully as Polanyi (1951: 157) puts it, there are two possible outcomes of unintended consequences analysis. There are both benign and malign unintended consequences. That is to say that because an order has arisen as the result of a process of unintended consequences (or spontaneously if you prefer) this does not mean that it is necessarily a 'good' or socially beneficial phenomenon. A further argument is required to explain why a system (such as the market) that operates through a medium of unintended consequences produces beneficial results. It is not enough to say that the origins of an order are the result of unintended consequences as this must be qualified by a separate or related contention as to the superiority of those spontaneous orders produced by unintended consequences.

Spontaneous order

The term spontaneous order has appeared at various stages in the history of political thought. John Millar, and his friend Francis Jeffrey, both use the term 'spontaneous' to describe the theory of political development which Millar had laid out (Hamowy 1987: 30), and the term spontaneous order also appears in the writings of Comte, Spencer and Durkheim (Klein 1997: 323). In these cases though either the term spontaneous was not accompanied by the term order, or the concept to which the term was applied was not the same as that with which we are concerned. It would appear that the first use of the term spontaneous order, as applied to the type of approach with which we are concerned, is to be found in Michael Polanyi's 1951 essay *Manageability of Social Tasks* (Jacobs 1998: 19).[19]

In his book *The Logic of Liberty*, Polanyi begins to deploy the term spontaneous order in the distinct sense in which we are interested. A spontaneous order is contrasted with a 'corporate order', the distinction being that 'there are certain tasks "which if manageable can only be performed by spontaneous mutual adjustments", tasks no corporate order is equipped to under-

take' (Jacobs 1998: 18). The key notion in Polanyi's discussion of spontaneous order is that such orders arise internally, that is to say the order is not
imposed by some external agency, but rather represents the formation of an
equilibrium by the mutual adjustment of individual 'particles' (Polanyi
1951: 155) in reaction to their surroundings – the settling of water in a jug
is a good example. Polanyi, and later Hayek, prefer to use the English term
spontaneous to describe this process rather than the term, credited to Köhler
by Polanyi, of a 'dynamic order' (Polanyi 1951: 154).

Spontaneity then is a qualifier of a type of order, and as such it is distinct
from an order qualified as 'externally imposed'. This difference, which
Hayek refers to as that between exogenous and endogenous orders (LLL vol.
1: 36–7), is the reason why the term spontaneous was chosen by Hayek and
Polanyi.[20] Having said this, and although the term spontaneous was adopted
in order to distinguish this species of order from types of exogenous imposed
order, Hayek would later admit that the word is ambiguous (LLL vol. 1:
viii–ix), and instead offered the terms 'self-generating order' and 'self-
organizing structures'. But for our purposes we shall proceed with the term
spontaneous if only because it has become the established critical term
applied to this tradition of thought. The other word in our key term is
'order', and here it is perhaps best to proceed with the technical definition
given by Hayek. Hayek describes an order as:

> a state of affairs in which a multiplicity of elements of various kinds are
> so related to each other that we may learn from our acquaintance with
> some spatial or temporal part of the whole to form correct expectations
> concerning the rest, or at least expectations which have a good chance of
> proving correct.
>
> (LLL vol. 1: 36)

This definition of order as a regularity owes a great deal to Hume's ideas
about the need for stability of expectations, and the focus is clearly upon the
internal nature of the order and not with its origins. That is to say order is
not considered as a command, as something that is by definition imposed,
rather Hayek's definition allows an order to be either the result of an external design (exogenous), but equally it may prove to be the result of a spontaneous adjustment (endogenous). The key characteristic of order is that it
provides some stability of expectation.

Once we have seen this it becomes clear that both Polanyi and Hayek
adopt the term spontaneous order in an attempt to contrast it with some
other form of order in such a way as to express the difference between exogenous ordering and endogenous ordering. This contrast, Hayek believes,
brings out the fact that there are many who cannot conceive of an order
which has not been deliberately established. It is precisely this misconception which Hayek seeks to oppose. We have then the term spontaneous
order which refers to a body of explanatory theory concerned with orders

which arise endogenously; and which its exponents contrast with orders which are externally imposed.[21]

Evolution

This is a term which frequently occurs in the writings of spontaneous order theorists, and it is worth noting two preliminary points. First, evolution, as deployed by Hayek and others, is not a technical term, nor is it strictly related to the use of the term in Darwinian biology. Rather the term is picked up by spontaneous order theorists as a contrast to notions of design and deliberate reformation. That is to say, rather than revolutionary change, spontaneous order theorists talk of evolutionary reform. The purpose being to emphasize the gradual, cumulative nature of changes which they ascribe to the unintended consequences of social action: to this extent spontaneous orders are said to have evolved. The second point follows on from our caveat concerning unintended consequences. Evolution is a descriptive term, it accounts for what exists and explains it in terms of the gradual development of social phenomena. Viewing evolution as a term that refers to the process of social change allows us to see that, like unintended consequences, it is a neutral term which is in no way prescriptive of beneficial outcomes.[22]

The invisible hand

The term invisible hand is perhaps the most famous description applied to spontaneous order arguments: it has, in Nozick's words, 'a certain lovely quality' about it (Nozick 1974: 18). The actual phrase occurs on three occasions in the writings of Adam Smith. Emma Rothschild (1994, 2001) has argued that each of these appearances is dissimilar, and this leads her to believe that Smith did not take the term at all seriously, but rather deployed it as an ironic literary device. If this is true then the use of the term invisible hand to describe spontaneous order theories is a misguided appropriation of terminology from Smith.

Smith's first use of the term is in his *History of Astronomy*, where he refers to the 'invisible hand of Jupiter' (EPS: 49) in connection with the errors of polytheism. Smith begins this section by observing that 'Mankind, in the first ages of society . . . have little curiosity to find out those hidden chains of events which bind together the seemingly disjointed appearances of nature' (EPS: 48). Thus he argues, in a crude society, irregular events are explained in terms of the intervention of an anthropomorphic Deity whose whims direct unexpected occurrences. The 'invisible hand of Jupiter' is not seen to act in the normal course of nature, but rather explains events which run contrary to its expected course (Macfie 1971: 595). Following Macfie we will argue that this early use of the term invisible hand does indeed differ from its later two appearances in the *Wealth of Nations* and the *Theory of Moral Sentiments*. Macfie argues that the later two uses refer to an invisible hand which restores the natural order, while the Jupiter example refers to a disruption of the natural order (Macfie 1971: 595). He notes that the Jupiter example is prior to the other two uses and that Smith, as a writer, recalled the 'pithy phrase' (Macfie 1971: 598) referring to the action of the Deity and applied it again in reference to an opposite conception of the role of the Deity. Thus for Macfie there is a conceptual link in that all three references are to the intervention of a Deity in human affairs, the difference being that the Christian Deity acts in the opposite manner to Jupiter (Macfie 1971: 595–6).[23] If we accept Macfie's thesis that Smith's later two uses of the term invisible hand differ from the first, in that they refer to differing conceptions of the Deity, then we are able to trace explicit similarities between the two later uses of the term which provide a conception of the invisible hand in the Christian era.

Smith's other two, more famous, uses of the invisible hand occur in his later works the *Wealth of Nations* and the *Theory of Moral Sentiments*. As these two passages will be discussed in some detail in Chapter 5, it will suffice here to assert that they are sufficiently conceptually similar to allow us to deploy the term invisible hand as a part of our conceptual vocabulary. The first, in chronological terms, of these two uses is that in the *Theory of Moral Sentiments*. In a section dealing with the effect of utility on the conception of beauty Smith discusses how distributive justice is played out in a commercial society (TMS: 184–5). He argues that the rich in a society are subject to the same physical constraints as the poor, that their corporeal frames restrict the amount which they can absolutely consume. As a result they are compelled to use their wealth to purchase the product of others' labour, and consequently they diffuse that wealth through society. Similarly, in the *Wealth of Nations*, the appearance of the invisible hand is again related to the co-ordination of self-interested action in order to produce benefits for the whole of society (WN: 456). Here the term invisible hand refers to the process, or mechanism, which brings about socially beneficial results from the interaction of self-interested actors. Whether that result is in the distribution of subsistence, or in the support of domestic industry, the process is the same: the whole of society benefits from the actions of individuals who did not have the good of society as their goal.

In both of these cases Smith uses the term invisible hand to refer to some imperceptible mechanism (the hand) which acts to produce benign results through the media of unintended consequences. The key factor here is that in each case the unintended consequences produce socially beneficial results. The mechanism of the invisible hand is that which creates benign spontaneous orders as the result of the co-ordination of the unintended consequences of human action.[24] Richard Bronk describes this concept, when applied to the economy, as a 'metaphor' (Bronk 1998: 92). Viewing the invisible hand as a metaphor for a particular mechanism of reconciling the unintended consequences of differently motivated actions allows us to use it as a technical term for the systems which produce socially beneficial spontaneous orders.[25]

Invisible hand arguments refer to the production of benign unintended consequences which manifest themselves as a spontaneous order. In order to use this conclusion to build a viable vocabulary we must oppose Robert Nozick's use of the term 'Invisible-Hand argument'. Nozick deploys the term to apply to arguments which we have separately categorized as spontaneous order arguments and unintended consequences arguments. That is to say he applies it to a variety of allied arguments that describe non-purposive interaction and the endogenous generation of order or patterns (Nozick 1974, 1994).[26] Nozick thus concludes that invisible hand arguments can produce both good and bad results (Nozick 1994: 192). Our understanding of this vocabulary is that unintended consequences arguments can produce benign and malign spontaneous orders, but that invisible hand arguments on the contrary involve an assertion as to the socially beneficial results of unintended consequences forming spontaneous orders.[27] The invisible hand refers to some mechanism which ensures the benign outcome of unintended consequence style arguments. The nature of this mechanism, and its relation to the concept of the evolution of knowledge will form the crux of this study.[28] In brief, unintended consequences arguments are a vital part of understanding the spontaneous generation of order in society (spontaneous orders), and those who argue in favour of this approach to social issues deploy invisible hand arguments when they wish to explain how the process of unintended consequences can produce benign results.

Or we might view it in this way. Unintended consequences arguments are one way of approaching social and historical processes, of this there is a subset concerned with the analysis of the formation of complex social orders which we have termed spontaneous order thought. Within spontaneous order analysis social change is viewed as an evolutionary process describing the gradual and cumulative nature of change in a neutral manner. Thus the results of this evolution may be viewed as either benign or malign orders. Benign orders are explained by invisible hand arguments. The invisible hand refers to some social mechanism, itself the product of evolution, which acts to produce socially beneficial outcomes from the interaction of actors pursuing their own purposes and operating under conditions of unintended consequences.[29]

2   The science of man

Science

Having laid out the tradition of spontaneous order and selected from among its exponents the groups identified as the Scots and the Moderns, exemplified by Adam Smith and Friedrich Hayek, we are now able to proceed in our search for the invisible hand. By analysing their explanatory approach to the social institutions of science, morality, law and government and the market we will begin to develop an understanding of the core elements of the approach. Our study views spontaneous order as an explanatory approach to social theory, and so it makes sense to begin our analysis with an examination of the Scots' views on science and social science.

The Enlightenment is often referred to as the 'age of reason', a time when a huge outpouring of learning and study existed against the backdrop of the first stirrings of an industrialized market economy, a time in some respects which paved the way for the modern world. In the area of science this was particularly true. The revolutions in scientific method which had shaped the natural sciences, from Bacon to Newton, had brought the consciousness of the nature and methodology of science to the forefront of academic enquiry. In the social sphere this new philosophy of empiricism, not new in the sense of never before practised, but new in the sense of being formulated and consciously undertaken, found a voice in the methodological writings of Newton and Locke. The Newtonian method so admired by the Scots rested on a desire to identify causal relationships from the observation of empirical data, verification by experimentation and then the formulation of simple and understandable general rules.[1] Newton's experimental method had led to great advances in the natural sciences which spurred on those who succeeded him to open up new areas of study. Thus Hume gives his *Treatise* the title 'A Treatise of Human Nature: Being an Attempt to introduce the experimental method of reasoning into Moral Subjects'. Hume is clear throughout his *Treatise* that reasoning, and explanation, must advance strictly in line with experience. He is explicitly rejecting a priori reasoning as a suitable method through which to advance scientific understanding. 'Observation and experience' (THN: 82) must be at the heart of the practice of science and, moreover, we ought not to extend our claims further than what is authorized and corroborated by experience.

Smith, in his *History of Astronomy*, expresses an understanding of science in terms of the human propensity to seek systematized knowledge. For Smith the purpose of science is explanation and the extension of knowledge, but this is not simply for the Baconian utilitarian reason that the knowledge of causes is power. Rather he explains the desire to practise science in terms of the sentiments.[2] Occurrences that disturb the course of our habitual expectations elicit in us a sense of 'surprise' (EPS: 40) at there having taken place. This initial surprise gives way to a sense of 'wonder' (EPS: 40) when we realize that we have nothing in our previous experience that can account for the event. Wonder is an emotion that strikes up a feeling of 'unease' (EPS: 36) within us, and the 'imagination feels a real difficulty in passing along two events which follow one another in an uncommon order' (EPS: 43). Wondrous events have this effect upon us, Smith believes, because of the manner in which we form our expectations. Our feelings towards events are shaped by our habitual acceptance of them and our expectation that they will continue to occur in the manner suggested to us by our previous experience. We develop habituated thought patterns or 'passages of thought' (EPS: 45) which 'by custom become quite smooth and easy' (EPS: 45) and we are shaken from this manner of approaching the world only by events which fail to fit into our established patterns of thought. It is in the reaction to such surprising and wondrous events that we are to find the original impetus to science. This 'psychological need' (Skinner 1974: 169) for the explanation of wondrous events leads us to seek understanding in terms of cause and effect.[3]

The impulse to explain, to calm the mind through understanding and ordering our thoughts is, for Smith, a facet of human nature which under-lines the gradual extension of the corpus of human knowledge. Once we have identified chains of cause and effect, and to some measure understood the relationships involved, it is as if a veil has been lifted from our eyes: 'Upon the clear discovery of a connecting chain of intermediate events, it [wonder] vanishes altogether. What obstructed the movement of the imagi-nation is then removed. Who wonders at the machinery of the opera-house who has once been admitted behind the scenes' (EPS: 42). Scientific enquiry does not rest simply with the dispelling of the initial sense of wonder. Once we have explained some part of the causal relationship our interest is piqued and we begin to enquire after other related relationships. Smith traces this point through his examination of the successive systems of astronomical thought. Again and again he stresses that the shift from an established theory to a new mode of thought is brought about by a 'gap' in the existing system. He writes: 'The imagination had no hold of this immaterial virtue, and could form no determinate idea of what it consisted in. The imagina-tion, indeed, felt a gap, or interval, betwixt the constant motion and the supposed inertness of the planets' (EPS: 91). This passage, referring to

Copernicus and Kepler, reveals the importance to Smith's theory of the progress of science of the notion of 'gaps'. A 'gap' for Smith may be an explicit hole in a theory, an area on which the theorist has nothing to say yet which is deemed vital to the coherence of his system; or it may be a weak point where the argument becomes over-stretched or over-complicated in order to explain a phenomenon within the terms of the theory. In either case a 'gap' is perceived which causes uneasiness through the theory's inability to lead the imagination along smoothly. This is the prompt to new enquiry and to the formulation of new theories.

Newton, according to Smith, was one such 'gap-plugger' whose system supplanted those that went before by building upon them and filling in the 'gaps' which troubled those who studied them. The strength of Newton's science lay not only in his strict methodological dependence on evidence and corroboration, but also on the fact that his system was, as a whole, more coherent, more understandable and thus more convincing than those which preceded it. This shows, for Smith, yet another aspect of the human propensity to seek systematized knowledge. Humans find beauty in systematized knowledge, and are thus naturally inclined, in order to ease the mind and to dispel wonder, to seek means to plug 'gaps' in their existing knowledge. The role of the philosopher for Smith is that of explaining phenomena in a coherent and 'gapless' fashion. But more than this, it is also a matter of providing explanations which are convincing; and this, as in the case of Newton, means not only that they are 'gap' free but that they are understandable to an audience. Thus when Smith wrote of Des Cartes' philosophy he referred to him as endeavouring 'to render familiar to the imagination' (EPS: 96) a series of difficult ideas through general rules.

The role of the philosopher is to provide 'some chain of intermediate events' (EPS: 44) whose coherence banishes wonder from the imaginations of mankind. Philosophy is thus 'the science of the connecting principles of nature' (EPS: 45), and as such it attempts to make sense of the universe and to calm the mind:

> Philosophy, by representing the invisible chains which bind together all these disjointed objects, endeavours to introduce order into this chaos of jarring and discordant appearances, to allay this tumult of the imagination, and to restore it, when it surveys the great revolutions of the universe, to that tone of tranquillity and composure, which is both most agreeable in itself, and most suitable to its nature.
>
> (EPS: 45–6)

As a result, philosophers, in order to calm our minds by the identification of the chains that bind the universe, are seekers after theory or 'systems' (EPS: 66).

Perhaps the clearest example of this is Smith's discussion of the early stages of science. The beginnings of the corpus of human knowledge lie, for

Smith, in the description and classification of phenomena. Our imagination seeks to 'arrange and methodize all its ideas, and to reduce them into proper classes and assortments', and as 'we further advance in knowledge and experience, the greater number of divisions and subdivisions of those Genera and Species we are both inclined and obliged to make' (EPS: 38). Thus we begin to form a system of knowledge based on the discrete classification of our experience. In other words we seek to order the world that we might better understand it, and thus calm our minds. Systems, in order to be attractive to us and to fulfil their role of plugging 'gaps' in our knowledge, must be explained and understood: a system must be coherent if it is to prove attractive to our sentiments.[4] There is, then, a real difficulty with systems which are over-complex; systems which are, in Smith's terms, 'of too intricate a nature to facilitate very much the effort of the imagination in conceiving it' (EPS: 89).

This, for Smith, is an explanation for the successive systems that he identifies in his *History of Astronomy*. Over-complexity, which arises when abstraction is required to explain phenomena within the terms of an established theory, lessens the hold of that system on the imagination. He argues:

> This system had now become as intricate and complex as those appearances themselves, which it had been invented to render uniform and coherent. The imagination, therefore, found itself but little relieved from the embarrassment, into which those appearances had thrown it, by so perplexed an account of things.
>
> (EPS: 59)

It is clear that, as human experience is not static, systems of knowledge must contrive to revise themselves in order to classify newly experienced events and phenomena. Should a problem arise here, if the system is unable to explain a phenomena, leaving a 'gap', or if the explanation is so convoluted as to fail to convince and allow the easy passage of the mind, then philosophers will begin a process of immanent criticism which will lead eventually to the development of a new system of thought (EPS: 71).[5] Smith's narrative of the shifts in the various systems of astronomical thought highlights this and ends, as we have seen, with an admiring survey of the work of Newton, lauding him for plugging 'gaps' left by previous systems. The desire to plug 'gaps' is vital to the success of systems of understanding. If, despite the system to which the imagination has become accustomed, we perceive 'two events [which] seem to stand at a distance from each other; it [the imagination] endeavours to bring them together, but they refuse to unite; and it feels, or imagines it feels, something like a gap or interval betwixt them' (EPS: 41). Then we become dissatisfied with our current system as its explanatory force fails and we are again left open to the unease of wonder.

Systems, in so far as they employ the language of cause and effect to explain phenomena, appear attractive to the human mind. Smith refers to systems of thought as being 'beautiful'. What he calls the 'beauty of order' (TMS: 185) is a sentimental reaction that links the ideas of utility and beauty. Humans come to develop a 'love of system' (TMS: 185). Not only do we become attached to particular systems of thought the more that they calm our minds, but we also become more attached to the habit of seeking after systems of thought.

What Smith has done in laying out his 'sentimental' theory of the origins of science is to downplay the primacy of utility, as advanced by Bacon, and the abstracted rationalism of the scholastic tradition. This is not to say, however, that Smith rejected all notion of utility as an original prompt to science, rather he downgraded its role in the early development of science and in the common impulse which prompts us to seek explanation. Both Smith and Hume devote some attention to the nature of utility as a prompt to action and they both draw on the same analysis to relate the two. They view utility as something that we regard as beautiful, that is, as something which affects our sentiments and prompts our admiration. Thus, while the decision as to the utility of an object is based on reason, our attachment to it is rather based on an emotional response to this usefulness (THN: 576–7; TMS: 179). We do not seek understanding in order that we might use it to our advantage: on the contrary we have an emotional need for understanding in order that our minds are able to function smoothly.

The desire to produce systems of explanation is a long-standing facet of human nature which has expressed itself in all our enquiries into the unknown and which, to a large degree, has shaped the course of the development of science. That thirst for knowledge which was so excited by Newton's advances in method is not a new attribute of human life, but is rather an aspect of human nature which has been slowly, yet gradually, developing for as long as humanity has existed. In Scotland this Newtonian influence and the natural propensity to seek explanation developed a particular focus: the Scots became determined to apply scientific principles to the study of man and society. Writers in the Scottish School became convinced that not only was it possible to apply scientific method to the study of man, but that it was vital to the foundational stability of the other sciences to do so.

Conjectural history

Of the Scots Hume is perhaps the most detailed in his explication of this issue. He is clear, in the introduction to the *Treatise*, that the science of man is vital to the progress of the other sciences. As he puts it: 'And as the science of man is the only solid foundation for the other sciences, so the only foundation we can give this science itself must be laid on experience and observation' (THN: xvi). Hume believed that all science depended ultimately on the science of man because all knowledge is based on human perception and learning. Thus, if we are to pursue accurate science to satisfy our emotionally driven need for explanation, we must seek to understand how humans understand. For Hume this meant, as we saw above, the application of the experimental method. This devotion to the experimental method, together with the aforementioned rejection of overly abstract theorizing, are the key characteristics of the Scottish thinkers' pursuit of the study of society. Science for the Scots was the careful study of the causal relationships by the means of which our imagination makes sense of the universe. Indeed much of the early parts of Hume's *Treatise* are given over to a careful examination of the relationship of cause and effect. Though we need not go into specifics here, it is perhaps wise to note Hume's conclusions on this matter as they were to prove so influential on many of the other Scots, in particular Adam Smith.

Hume believed that the body of human knowledge is based on a series of habitual associations, garnered from experience, which exist in the imagination. Notions of cause and effect allow the mind to pass easily from one phenomenon to another by the steady flow of the imagination. Thus when we see a flame we expect to feel heat, because we have always done so. Our experience allows us to infer, or leads us to expect, phenomena to occur in that order in which we have previously known them to arise. The attribution of a causal relationship, from flame to heat, is based on a series of key conditions which, for Hume, build a causal relation in our minds drawn from our past experience. The principles of contiguity, of succession and of constant conjunction – that phenomena are somehow connected in situation, that one succeeds the other, and that they are always, in our experience, found in this manner – leads us to presume some necessary connection between them. The path of our mind, drawing on past experience, leads us to infer, or habitually to expect, phenomena to occur in the manner in which they have always done. For Hume this process does not demand that we have an understanding of the precise nature of the causal relationship. We need not have any perception or impression of why the two phenomena are related: it is enough thus far that we 'know' them by habit to be related. The precise identification of the causal relationship, or 'necessary connection', is not necessary for us to accept that such a relation exists. Hume believed that it was a far more profitable approach to follow the habitual nature of our attribution of cause and effect, and to seek out causal regularities from which we may then proceed to our enquiries rather than to proceed by the formulation of abstracted hypotheses. The fact that we cannot identify the nature of a causal relationship, or in some cases that we cannot from our current experience find a cause to relate to the effect of a specific phenomena, does not for Hume entail that we should abandon our enquiries and attribute that phenomena to chance. Attributing things to chance cuts off all further hope of understanding, so for Hume what is explained by reference to chance is nothing but 'secret and unknown causes' (EMPL: 112), or causal relationships which we have yet to explain from experience.[6]

If our knowledge is based on a series of habitual relations grounded in experience, then this further supports the pursuit of an experimental approach to scientific enquiry. This empirical method is an attempt to formalize and to understand the habitual acceptance of causal relations which allow us to carry on our lives. The chief difficulty here for the Scots is the same problem which has plagued the social sciences since their inception: that there is little or no scope for controlled experimentation. Hume realizes this early on: in the introduction to the *Treatise* he describes how moral philosophy has a 'peculiar disadvantage' in that 'premeditation' by the actors will 'disturb' the validity of the observations (THN: xviii–ix). The Scots, then, must find some other way of adapting the experimental method to social situations, and, while doing so, be able to retain all that makes this method so successful.

In answer to this problem the Scots hit upon the extension of the observational method to human life as it is, and as it has been lived. If the natural scientist draws on his observation and experience to build a theory which is supported by careful observation of controlled experiments; then the Scots, in the absence of controlled experiments, would simply extend the observational and experimental bases and seek corroboration for their thought in the records of human experience which form history. As Hume puts it:

> We must therefore glean up our experiments in this science from a cautious observation of human life, and take them as they appear in the common course of the world, by men's behaviour in company, in affairs, and in their pleasures. Where experiments of this kind are judiciously collected and compared, we may hope to establish on them a science, which will not be inferior in certainty, and will be much superior in utility to any other of human comprehension.
>
> (THN: xix)

This gives rise to the Scots' peculiar focus on, and fascination with, historical writings, and to their own specific comparative approach to the study of history – the common pursuit of which is to be found in the work of the whole school of thinkers. It is this 'conjectural history' that marks out the Scots in the early development of social science. Dugald Stewart is clear that the development of what he calls 'Theoretical or Conjectural History' (Stewart 1793: 293) was one of the key accomplishments of Smith and the Scottish Enlightenment.

In many respects the Scots can be regarded as among the first practitioners of the comparative method which has become so central to social science.[7] Hume stresses the vital importance of comparison to our mode of thinking. He argues that all 'reasoning' is a form of 'comparison' (THN: 73), and that our judgements are all more or less based on comparisons between phenomena. From this Hume is able to develop an argument that our judgements are based on comparison, that is to say that we approach objects not in relation to their intrinsic value, but rather in relation to their subjective, comparative value (THN: 372).[8] He writes in his essay *On the Dignity or Meanness of Human Nature* that our moral judgements are 'commonly more' (EMPL: 81) influenced by processes of comparison than by eternally fixed standards.[9] The Scots believed that the examination of corroborated evidence would allow them to draw out rules of behaviour that would amount to a scientific history.

Such a method is based on a number of key assumptions. The most important of these is that there exists a universal human nature which is, at base, unchanging. The habitual expectations which form the basis of natural science are premised on the notion that nature is in some basic sense unchanging; we expect there to be heat with a flame because it has always been so (Hume's constant conjunction). In the social sphere this model of constant conjunction must be present to allow us to form habitual expectations. Thus there must exist some universal principles of human behaviour, and for the Scots these lay in the principles of human nature. These principles of human nature are themselves to be drawn from the careful observation of historical experience through the comparative method. A conjectural history of human nature is what underpins the broader practice of conjectural history.[10] Drawing on what we know of ourselves and our own actions, and corroborating this with examples from history and literature allows us to identify key aspects of human behaviour that are in some sense transhistorical. That is, they are akin in form to the universality of nature, or that they are as much natural laws as those which guide the physical sciences.

If the evidence of history and observation may be drawn upon to show regularities in human reactions to external stimuli, then we are able to formulate some simple principles (or general rules) that apply in all ages and in all cultures. As Hume famously put it:

> It is universally acknowledged that there is a great uniformity among the actions of men, in all nations and ages, and that human nature remains still the same, in its principles and operations ... Mankind are so much the same, in all times and places, that history informs us of nothing new or strange in this particular. Its chief use is only to discover the constant and universal principles of human nature by showing men in all varieties of circumstances and situations and furnishing us with materials from which we may form our observations and become acquainted with the regular springs of human action and behaviour.
>
> (ENQ: 83)

The Scots' conception of a universal human nature was not, however, a species of crude ahistoricism or cultural insensitivity.[11] On the contrary the Scots made it the centre of their study to examine the differences which arise in human actions through time and across cultures. Far from seeing these differences as aberrations from a universal human nature that indicate a weakness of the Scots' approach, Hume instead drew upon them to deepen the Scots' project. He argued: 'All birds of the same species in every age and country, build their nests alike: In this we see the force of instinct. Men, in different times and places, frame their houses differently: Here we perceive the influence of reason and custom' (ENQ: 202). What this illustrates is that the condition of society affects the broad manifestation of human nature.[12] That is, the manifestation of the human need for shelter differs through the influence of experience and custom: that is the context is important.[13] But beyond this, in keeping with the desire to seek out a few simple rules for the science of man, the Scots held that there was an underlying universality in the fact that humans do indeed always seek shelter and that this underlying universality could be identified from the careful observation of history. In Hume's assertion we see a distinct claim that the explanatory factor for the difference of behaviour in different places is 'reason and custom', that is to say that social or 'moral' causes explain the differences in human behaviour. Here Hume is asserting a view common to all of the Scots and vital to their project of a scientific history. For if 'physical causes' (EMPL: 198) are held to determine behaviour then the idea of a universal human history is exploded into that of discrete geographical histories. The Scots must reject any sense of the determination of human behaviour being explained by the physical environment if they are to be able to conduct conjectural history on a grand scale to support the science of man. In his essay *On National Character* Hume demolishes the argument from physical causes most famously associated with Montesquieu. What Hume argues is that the customs and behaviour of humans cannot be attributed to the physical environment alone. Elsewhere he goes so far as to argue that moral causes are of equal strength to physical causes in his famous example of the prison (THN: 406). Here a man is kept in prison as much by the moral cause of the guard's incorruptibility as by the stone of the walls.

Hume's argument against physical causes hinges on two important historical facts. First, the manners of a people differ through time while their physical environment remains the same. Thus the inhabitants of Germany described by Tacitus differ vastly from the modern inhabitants of the same physical environment: climate cannot be the deciding factor in the explanation of this change. Similarly the customs of near neighbouring countries, such as England and Scotland, differ to an extent that cannot be explained by the small difference in their physical situation. The Scots must explain the change in the customary practice of the inhabitants of a particular situation through time. To achieve this they draw strongly on their conception of moral causes. Through conjectural history the Scots developed a 'four stages' theory of cultural change which accounted for difference through time by reference to the changing economic condition of the country. This, they believed, was a more convincing explanation than mere physical situation. This is not to say, however, that the Scots totally rejected the significance of physical causes, rather they viewed their influence as operating through moral causes. As Dunbar puts it: 'causes physical in their nature, are often moral only in their operation' (Dunbar 1995: 296). The influence of the physical environment is filtered through the medium of moral, particularly economic, causes.

The Scots' social science was an attempt to examine and to explain these differing customs. By comparing, for example, the records of Tacitus on the Germans with those of contemporary writers on the Native Americans – as Robertson (1769) did – it is possible to corroborate each account by noticing the similar characteristics.[14] From this we are able to identify common features which can be attributed to human societies in a particular stage of development. This allows the Scots to produce stadial theories of history and Smith, Millar, Kames and Ferguson all draw on this to explain the customary practices of societies at differing stages of development. Conjectural history must, however, be undertaken with care and scientific rigour. All sources must be considered for their accuracy and veracity, and the observations which they recount must be corroborated by like evidence from other sources, both historical and from our own experience of human behaviour. Before building a theory around the evidence of conjectural history the Scots wanted to be particularly sure that the evidence with which they dealt was genuine. This having been established by the careful selection and comparison of sources, the Scots were then free to search for causal regularities and to form theories as to the determining factors in the development of human society.

Simple models of understanding

Before proceeding to an examination of the Scots' rejection of what they view as mistaken approaches to the explanation of social phenomena, we ought to note that many of the problems that they identify with what we will call 'simple models of explanation' are a direct product of the human desire for order and explanation. There is a range of problems that the Scots associate with this human love of systems. Chief among these is what Smith refers to as the 'spirit of system':

> From a certain spirit of system, however, from a certain love of art and contrivance, we sometimes seem to value the means before the end, and to be eager to promote the happiness of our fellow-creatures, rather from a view to perfect and promote a certain beautiful and orderly system, than from any immediate sense or feeling of what they either suffer or enjoy.
>
> (TMS: 185)

There is a danger that, in our desire to hold onto the systems that we create, we run the risk of losing our sense of priorities: of elevating the means above the end.

It is possible that our attraction to systems may lead us to over-extend or over-simplify them in order to keep hold of them. Once we hit upon a system of thought that seems to serve us well in some areas, we are prone to extend its terms of reference beyond that area in which it first arose. Smith makes this point when he notes:

> systems which have universally owed their origin to the lubrications of those who were acquainted with one art, but ignorant of the other; who therefore explained to themselves the phaenomena, in that which was strange to them, by those in that which was familiar
>
> (EPS: 47)

Some philosophers tend to excessive abstraction from the original bases of their thought, leading to systems of thought which fail to convince the imagination and lead us to the confused apprehension of 'gaps'. But beyond this systems themselves are open to the same errors as the practice of all science. A philosopher who builds a system on selective or limited grounds is open to the possibility of extending their system through sheer devotion to it, of extending it even when faced by contrary evidence. Thus it becomes a real danger that the system itself may become too abstract to fulfil the calming role that it was originally conceived to fill. And in the end we might end up deceiving ourselves in order to preserve a system in which 'gaps' have become apparent through over-complexity.

This is particularly the case in moral systems where the Scots identify an inclination in some which leads to factionalism and the development of sects which hold a devotional belief in a certain system of moral thought. There is a real problem here: our attraction to a particular system of belief often becomes such that we view it as infallible, our belief in it becomes akin to a faith. Thus Hume attacks: 'all moralists, whose judgement is . . . perverted by a strict adherence to system' (THN: 609). This problem is further com-pounded in the moral sphere by what the Scots identify as problems in the creation of systems. Systems of moral thought are often the product of party or factional bias, they are created to support political positions rather than to advance understanding (EMPL: 160; TMS: 232).

As Hume notes in his essay *Of the Original Contract*, the contrasting models of contract and divine right guided the factional struggles between Whigs and Tories, Protestants and Catholics, and Hanoverians and Jacobites in the party politics of seventeenth- and eighteenth-century Britain (EMPL: 465). Hume (EMPL: 64) and Smith (LJP: 402) both argue that these con-trasting theories were the creations of those factions which advanced them. Smith states that the natural inclination towards, or the emotional attach-ment to, Whig or Tory positions predates either theory and that the theories themselves were in turn developed to support pre-existing political posi-tions. The Scots were instantly wary of systems built around factional positions and not drawn from scientific examination of the evidence.[15] As a result they reject both divine right and social contracts as the historically inaccurate product of partisan politics. Appeals to God or the social contract are attempts to justify a particular political system and not a scientific attempt to explain the development of political systems.

Arguing from the historical record and from the comparative evidence of recently discovered primitive societies such as the American Indians, Ferguson forcefully makes the point that there is no historical evidence to back up the sorts of states of nature which have been used as the basis for contract theories. Indeed Ferguson argues that such states of nature are fictions created by writers in support of their own theories (ECS: 8).[16] If there is no record or evidence of such a state of nature having existed, or of a contract having been the means of humanity's exit from it, and, more importantly, if 'both the earliest and latest accounts collected from every quarter of the earth, represent mankind as assembled in troops and companies' (ECS: 9), then it is clearly a nonsense to have a state of nature and an original contract as the basis of a complex political theory through which current political establishments are viewed or justified and the origin of society is explained. Ferguson goes further than this, he seeks to define humans in social terms, thus what we understand by a human, is a creature who has always existed in society with others and whose attributes are framed by this fact (ECS: 23).

Ferguson is clear that such tales of a state of nature are precisely that: that as a result of the human desire to systematize knowledge and to explain the unknown, humans create simplifying myths which they then use to plug the 'gaps' in established knowledge. For Ferguson humans are by nature sociable, and therefore no contract is required to bring them into society. The sociability that is the basis of human society is a facet of human nature, a fact to which all the evidence of history and experience points. The unfolding of human actions and abilities is the true state of nature, and this means that every condition in which we find them is equally natural. As Ferguson puts it:

> If we are asked therefore, Where the state of nature is to be found? we may answer, It is here; and it matters not whether we are understood to speak in the island of Great Britain, at the Cape of Good Hope, or the Straits of Magellan.
>
> (ECS: 14)[17]

There is a danger in moral philosophy when the tendency to factionalism renders moral debates little more than subjective arguments, but the Scots wish to go beyond this and to seek out objective truth in moral matters. They believe that moral philosophy, open as it is to error on so many grounds, must be undertaken with the utmost care and scientific rigour. We must be cautious, in examining moral matters, to keep our speculations in line with the evidence, and in seeking clarity of expression that allows a clear debate. At the same time, and in order to remain scientific in our approach, we must relate 'philosophy' to real life and to experience: our theorizing must be explanatory.[18]

We have already seen that the Scots had developed their own particular scientific methodology for the pursuit of social science and we have noted the key role to be played in this by their conception of a conjectural history. Before proceeding to an examination of the Scots' understanding of the origin and development of social institutions grounded in their conjectural history, it is perhaps best to examine their use of the 'scientific' method to oppose certain pre-existing models of explanation and understanding. As we have seen, the Scottish Enlightenment wished to set the study of society on secure methodological foundations akin to those which had recently taken hold in the natural sciences, and each thinker in turn rejects the established systems of political study in favour of a new, more 'scientific' in their eyes, understanding of the origins and legitimacy of society and government. For a proper understanding of the significance of the Scots' arguments against established models of political understanding it must first be noted that each writer argues against models which involve the direct intervention of the deity and against models of what we might call, after Hayek, 'construc- tivist rationalism'; and that in opposition to these views they advance arguments favouring the gradualist unintended consequence style of devel- opment which typifies the spontaneous order approach. The Scots' rejection of these established systems of understanding is premised on a belief in the unlikelihood, and in some cases the impossibility, of deliberate human action acting according to a prior, rationally conceived plan, having been responsible for the origins of society and government.

This desire to view society as a deliberately planned artifice is noted by both Smith and Ferguson. Ferguson argues against such systems of thought, contesting that they are simplifications based on the absence of record or knowledge. As he puts it:

> An author and a work, like cause and effect, are perpetually coupled together. This is the simplest form under which we can consider the establishment of nations: and we ascribe to previous design, what came to be known only by experience.
>
> (ECS: 120)

Such approaches are, for Ferguson, unscientific and at odds with the empiri- cal and scientific method that he wishes to pursue. Attempts to explain the origins of society which do not ground themselves in established evidence, or reasonable conjecture based on that evidence, fail, in Ferguson's view, to provide a firm basis for knowledge. Drawing on historical evidence and 'scientific' understanding the Scots set out to oppose forms of explanation which they regarded as simplistic, inaccurate, unscientific and thus mislead- ing. The first of these forms, that of direct divine intervention, is dealt with during the Scots' various writings on polytheism, superstition and miracles.

The Scots, undertaking a conjectural or 'natural' history of religion, generally refer to polytheistic religious belief as the earliest and most simple form of understanding. As a result it is, for the Scots, the belief system of the uninstructed savage. The savage, who lacks understanding because of the precariousness of their position – that is to say that their pursuit of subsistence narrows their attention to that one activity – is not in a position to reflect on the nature of the universe.[19] That which they do not understand, yet which attracts their attention, they explain in terms of the direct intervention of anthropomorphic deities. Thus the 'irregularities of nature' (EPS: 48) are attributed to the direct actions of a multiplicity of deities which each have power over their set field, and act from motivations analogous to those of humans. According to the Scots such beliefs were founded on ignorance: they were superstitions which sought to explain gaps in our knowledge and experience by attributing purposive actions to invisible deities. What savages could not explain from their limited experience they attributed to some invisible superhuman being which differed from humans not in its motivations but only in the extent of its powers.[20]

Such forms of understanding, the Scots believe, fall out of favour as society progresses economically. As material development occurs in line with the growth of experience-based knowledge, humans gradually acquire a degree of leisure time in which they are able to consider those irregular phenomena previously ascribed by them to specific deities. As enquiry grows, according to Smith, humanity becomes more favourable to ideas that reveal the links between phenomena. Humans therefore become 'less disposed' (EPS: 50) to notions of direct divine intervention and develop more complex understandings of phenomena. They begin, so Smith thinks, to view the phenomena of nature as regular rather than irregular occurrences, and thus begin to enquire into the nature of this regularity and the connections between phenomena in order to satisfy the human desire to produce systematic knowledge. Understandably this process is accelerated amongst those cultures that have achieved the greatest material advance, and within societies by those groups which have access to the greatest leisure time.

Polytheistic religion falls out of favour as society progresses in terms of material wealth and knowledge. The advent and pursuit of science hasten the demise of a belief in direct divine intervention and it gives way to forms of religion that view God as an agent of creation though not of everyday action. The creation myths of polytheistic religions, which explained the creation of their deities as well as of the world, give way to the notion of a single and universalistic Deity responsible for the 'complete machine' (EPS: 113) of all creation. Such a Deity becomes progressively refined as the 'natural history of religion' proceeds and is gradually stripped of anthropomorphic features and motivations. This process continues until, in the Scots' view, a position of Theism is reached whereby the Deity's actions proceed according to general rules which govern the whole of the universe, are observable in nature, and not to be attributed to the direct intervention of God. The Scots were deeply suspicious of cases where direct intervention, such as miracles, were claimed. Hume in particular rejects the notion of a miracle much in the same manner as he dislikes attributing phenomena to chance. Miraculous events for Hume are the product of 'secret and unknown causes' (EMPL: 112) which may or may not be divine in origin but which ought to be examined through the same rigorously scientific lens as all other natural phenomena. Similar superstitions, such as the belief in witches, will fall out of use as human knowledge and society advance, and a key part of that advance was the scientific approach to phenomena and the rejection of direct divine intervention.

Having rejected direct divine intervention as an explanatory model the Scots continue their 'demotion of purposive rationality' (Berry 1997: 39) by criticizing the classical doctrine of the great legislator which ascribes the institution of states and legal systems to the deliberate actions of great individuals or to direct human agency. The most detailed rejection of great legislator arguments in the work of the Scots is to be found in the writings of Ferguson and Millar.[21] Both of these writers utilize similar arguments against what they view as a product of mythological history, and instead advance their own 'scientific' understandings of the issues involved. Ferguson and Millar present two clear arguments for the inaccuracy of great legislator theories. Their first view is that there is no accurate and detailed historical record of such figures, and their second is that the very notion of one individual being able to organize the whole of a society is improbable. As Millar puts it:

> as the greater part of those heroes or sages that are reputed to have been the founders and modelers of states, are only recorded by uncertain tradition, or by fabulous history, we may be allowed to suspect that, from the obscurity in which they are placed, or from the admiration of distant posterity, their labours have been exaggerated, and misrepresented.
>
> (Millar 1990: 7)

Instead of accurate historical evidence what we have are mythical histories based on the human propensity for the assumption of design. Such semi-historical, semi-mythical figures as Lycurgus, Romulus, King Alfred and Brama have had the design of social systems attributed to them by popular histories which lacked the analytical tools to view the institution of laws by any other means than deliberate constitution writing.[22] Such simple models of understanding demonstrate the most obvious mental approach to filling 'gaps' in our knowledge. An institution exists, therefore it must have been created and the simplest model for understanding this is to attribute the artifice to a single artificer. Ferguson argues that the origins of society are obscure and come in an age before philosophy, which leaves us in a position of ignorance as to the origins of our social systems. Proper social science, conducted along scientific lines through conjectural history allows us to discount such belief systems by explaining them in terms of a primitive lack of awareness coupled with the natural propensity of human nature to seek explanation. We invent explanations that, in a savage condition, calm the mind, but which, as knowledge evolves and advances, no longer fulfil this role: so undeveloped societies attribute to this model of author and creation institutions which instead arose spontaneously, through the medium of unintended consequences.

To underline this argument Millar brings into play the nature of life in primitive societies. That is to say that the immediacy of savages, their limited attention to their immediate physical needs (Millar 1990: 3), makes it unlikely that a savage society would produce one individual capable of the conscious creation of an entire system of law.[23] Leading on from this point, several of the Scots point out that even if such an individual had existed, they would have faced great difficulties in enforcing their projections on a people whose attention was consumed by their immediate needs. Aside from the difficulty of enacting such plans both Ferguson (ECS: 119) and Millar (1990: 7) imply that long-term rational planning is not an accurate manner in which to depict the everyday organization of 'law' in early societies. Hume argues that in the early ages of society the task of setting down laws is so complex that in itself it prevents the possibility of one individual shaping the whole system, he writes: 'To balance a large state or society ... on general laws, is a work of so great difficulty, that no human genius, however comprehensive, is able, by the mere dint of reason and reflection, to effect it' (EMPL: 124). Having said this the Scots do not go so far as to completely deny the existence of great legislators, instead they argued against viewing such individuals as the sole authors of complete systems of law.[24] Great individuals have existed and been associated with reforms of the legal system – decent historical evidence of the more recent figures such as Alfred exists – but their role in this is far from that mythologized in popular history. Indeed the Scots point out that the very universality of the institution of civil government, and the similarities between forms of government in different places and ages strongly suggests that their origins cannot lie in the work of single individuals.

Though they argue that the prevalence of government as an institution militates against its origins having lain in the genius of particular individuals (ECS: 121; EMPL: 275), they nonetheless accept that the diversity of forms of government both between countries and within the same country at different times, is an important issue which requires explanation. What the Scots hit on to explain this diversity is a development of their view regarding the human propensity to systematize. Great legislator myths, as we have observed with polytheism, express this facet of human nature; the desire to link cause and effect in a direct manner is the most simplistic form of human understanding. It is, however, just that, too simplistic. Hume argues in detail about the nature of cause and effect in the *Treatise*, but it is in his essay *On the Rise and Progress of the Arts and Sciences* where he applies it to traditional views of political foundations. Here Hume argues that the complex situations and influences which determine those causes which act upon a vast body of people, such as a society, are less subject to the casual interference of one particular incident. That is to say that the forces which shape a society, the causes if you like, are far more complex while at the same time being far less subject to the control or interference of any one particular person. A further corollary of this is that these forces are thus more easily understood through the headings of general descriptive laws, rather than through any notion of direct agency (EMPL: 112). Understanding of the origins of society ought not to progress as though there were a creator and creations, a direct cause and effect, but instead must seek to proceed by the formulation of general rules from the observation of events.

If complex social institutions are better understood as the product of the interaction of many individuals within a specific context, then we are able to see what the Scots set up as an opposing model of the understanding of the development of law and government. For the Scots it is the reaction to particular circumstances which shapes a legal system and not conscious planning: 'The croud of mankind, are directed in their establishments and measures, by the circumstances in which they are placed; and seldom are turned from their way, to follow the plan of any single projector' (ECS: 119). Thus, even when legal codes are consciously set up, such as those attributed to Lycurgus or the Twelve Tables at Rome, these are themselves the product of immediate concerns and knowledge filtered through the medium of popular opinion, and not the result of a rational attempt to design the entire legal system. Approaching the origin and development of law in this manner emphasizes the importance of customs. Custom plays a prominent role in the Scots' analysis along with its related concept of habit: social practices that are found to be useful become habituated and continue as customary behaviour which is refined through use and as circumstances arise. For this reason the Scots argue that custom and habit inhibit the introduction and enforcement of planned legal systems and constrain the action of legislators. As Millar would have it:

> it is extremely probable, that those patriotic statesmen, whose existence is well ascertained, and whose laws have been justly celebrated, were at great pains to accommodate their regulations to the situation of the people ... and that, instead of being actuated by a projecting spirit ... to produce any violent reformation, they confined themselves to such moderate improvements as, by deviating little from former usage, were in some measure supported by experience, and coincided with the prevailing opinions of the country.
>
> (Millar 1990: 7)[25]

The key to understanding the process of law formation and the role played within it by specific legislators, lies in the concept of the socialization of these legislators. The Scots take great pains to stress that the origin of law lies in custom and established opinion. Thus even where great legislators have played a role in framing the codified version of laws, they have merely acted in line with the popular sentiments of their day, and have not proceeded from some original and metaphysical plan. Several of the Scots make this point explicitly. Ferguson argues that such statesmen 'only acted a superior part among numbers who were disposed to the same institutions' and that their fame credits them as 'the inventors of many practices which had been already in use, and which helped to form their own manners and genius, as well as those of their countrymen' (ECS: 121).

The thrust of the argument is not simply that a legislator would craft his laws to find favour with the established views of the people in order that he would be able to enforce them, but it is a broader argument about the origins of law and indeed knowledge. Both law and knowledge are the product of the socialization of individuals within the particular traditions and customs of their country. Laws are the product of the interaction of people that merely find articulation in the specific enactments which become associated with particular legislators. Law exists as habit, custom and opinion before it can be consciously codified into written form. What the Scots are doing here is downplaying the significance of the role played by the deliberate action of individuals in social processes in favour of a description which stresses the socialized generation of values and institutions.

For this reason the simplifying myths of great legislators are wholly inaccurate as 'scientific' understandings of the origin and development of law. Law and legal reform are a gradual evolutionary process, not the result of a concerted plan. The Scots instead refer to the example of experience inherent in habitual and customary practice and propose gradual and incremental reform in line with existing knowledge and in reaction to changes in circumstances as the model through which to understand the development of social institutions.

Through the chapter we have seen that the Scots sought to provide an explanation of the impetus to and nature of science. They grounded this explanation in what they saw as a universal characteristic of human nature: the emotional need for order and explanation to stabilize expectations. This leads to a human propensity to classify experience under generalized rules and categories. Cause and effect are understood as habitual relations based on experience, and explanations invoking cause and effect (scientific explanations) represent a formalization of the mental habit of classification that constitutes the human mind. The body of scientific thought evolves through a process of 'gap-plugging' to provide increasingly satisfactory and coherent explanations. In the social sciences the chief source of empirical evidence is history, and through conjectural or theoretical history the Scots form an explanatory social theory that accounts for the existence of human institutions. Drawing on comparative analysis, they develop a composite model of social development based on an underlying universal conception of human nature that interacts with different circumstances to produce social diversity.

The Scots rejected a series of pre-existing explanatory models on the ground that they are un-scientific over-simplifications. By conducting a 'natural', or conjectural, history of religion the Scots demonstrate that the belief in direct divine intervention is an 'uninstructed' belief that calms the mind of those in less developed states, but which increasingly fails to satisfy the imagination as society advances. Similarly, they argue that the notion of a great legislator or founder of a state is a simple model of explanation that plugs a gap in human knowledge regarding the origins of political societies. They trace this simplification to a propensity to view that which exhibits order as having been deliberately designed. The Scots reject this view as implausible given the circumstances of the societies in which government first arose. Political establishments are the result of a gradual reaction to circumstances: they do not represent a planned or deliberately patterned system. Even when legislators have acted they have done so in line with opinion and, more significantly, they have acted as individuals socialized within the cultural traditions of the group. They also reject the social contract approach, arguing that there was no state of nature and no contract. There is no need to seek to understand the origins of political establishments through a contract model: for, if we can provide a 'scientific' explanation of the origins of political societies, then we have no need of an alternative hypothetical justificatory model. The Scots viewed these existing models as simplistic rationalizations that fail to proceed in a properly scientific manner and, as a result, fail to provide satisfactory explanations.

3 The science of morals

Sociability

Having laid the ground for their rejection of established theories of the origin of social institutions the Scots are then free to develop their own understanding of how such institutions came about. As intimated in the previous chapters, this understanding is strongly grounded in notions of habit. The Scots regard habit as a universal attribute of human nature. Ferguson defines a habit as 'the acquired relation of a person to the state in which he has repeatedly been' (Ferguson 1973 vol. 1: 209), while Hume argues, with relation to the habitual behaviour of all animals, that what is commonly referred to as instinct, can in fact be understood as a form of habitual behaviour. Indeed he regards habit as 'one of the principles of nature' (THN: 179).

In terms of mankind, the Scots' understanding of human psychology and the nature of science itself are, as we have seen, stated in a language of habit and habitual relations. The Scots' 'sentimental' theory of psychology draws strongly on notions of habit.[1] Hume argues that 'the far greatest part of our reasonings . . . can be derived from nothing but custom and habit' (THN: 118), and, as we saw above, his ideas about the customary transition of ideas in the imagination, and of constant conjunction, treat our mental faculties as highly subject to the force of habit. Reasoning in terms of cause and effect was, for Hume, a species of mental habit grounding our expectations which are drawn from past experience.[2] Thus belief and reasoning are a customary process. We become habitually accustomed to a particular chain of events and come to form expectations regarding them.

A habit of behaviour is acquired in the same manner as a mental habit in the Scots' analysis. Habits are acquired through practise, yielding a constant conjunction in the mind; and this conjunction is strengthened, growing 'more and more rivetted and confirmed' (EPS: 41), through repetition. We come to accept a habit (or custom) because it falls in with our experience of the world and as such acts to calm the mind. Habit is one of the strongest forces that attaches humans to certain practices or modes of thought. Smith follows this line of argument and believes that the strength of habit is such that it shapes our emotional responses to external phenomena, making us 'used' to certain things. He writes:

> It is well known that custom deadens the vivacity of both pain and pleasure, abates the grief we should feel for the one, and weakens the joy we should derive from the other. The pain is supported without agony, and the pleasure enjoyed without rapture: because custom and the frequent repetition of any object comes at last to form and bend the mind or organ to that habitual mood and disposition which fits them to receive its impression, without undergoing any very violent change.
>
> (EPS: 37)

Habits, drawn from experience, act as a non-deliberative guide to our behaviour, they allow us to form expectations around which we are able to order our actions.[3] One example Hume gives of this is when we hear a voice in the dark we suppose that someone is near us even though it is only 'custom' that leads us to believe so (THN: 225).[4] Such habitual inferences provide us with a degree of stability which is sufficient for us to carry on our lives (THN: 64). We draw on our experience to form mental habits of cause and effect that we are then able to use to stabilize our expectations and to reduce uncertainty in the mind. This stabilization of expectations by habituation is extended into our relations with others who also operate on a like model of understanding.

The Scots commence their conjectural history of social phenomena by building on their rejection of a state of nature that existed prior to the formation of society. Humans, the Scots contend, 'are to be taken in groupes, as they have always subsisted' (ECS: 10).[5] The universality of this observation leads to the conclusion that humans are by nature sociable, that human nature is social.[6] But the Scots' analysis is more sophisticated than this simple assertion: they go on to examine the dynamics of sociability which develop around the 'habit of society' (ECS: 11). What the Scots seek to explain is the reason why humans are always found in groups, the reason why they are sociable: in other words they seek to discover what it is that binds humans together. Their answer to this question as to what binds society together is to be found in a complex interrelation of the concepts of utility and sympathy.

Sociable individuals come to form unwritten or non-deliberatively generated conventions of behaviour – in Hume's famous example two men rowing a boat 'do it by an agreement or convention, tho' they have never given promises to each other' (THN: 490) – which develop into customary modes of behaviour as we repeat them and see the utility in them.[7] The convention that arises is a spontaneous order that is an unintended consequence of an adaptation to circumstances.

Smith also describes a like process in the *Theory of Moral Sentiments*. He begins from the now familiar notion of the 'habitual arrangement of our ideas' (TMS: 194) and proceeds to note that our notions of taste are in a large part formed by the influence of custom and habit (TMS: 194–6). He goes on to examine the diversity of customs that have arisen, along the way distinguishing between a custom and a fashion, the latter being more transient and of a weaker influence upon the sentiments (TMS: 194–9).

Discussing the aesthetics of Pope, Smith notes his view that 'The whole charm' of our notions of taste 'would thus seem to arise from its falling in with habits which custom had impressed upon the imagination' (TMS: 199). Smith though believes that this is not sufficient. For if all of our judgements of beauty were made with reference to past habits, then no innovation would be possible. On the contrary Smith believed that humans were highly attracted to new phenomena and that these, far from depending solely on fitting in with established tastes, drew their beauty from their utility; a principle which acts independently of custom (TMS: 199).[8]

As we become familiar with an innovation whose utility is apparent to us, we absorb it into our habitual practice. 'Custom has rendered it habitual' (TMS: 201) to us and we draw on our experience of it habitually and non-deliberatively rather than through constant reference to its utility. Custom shapes our behaviour and our expectations of the behaviour of others. We judge others' behaviour according to how it fits our habitually formed expectation of what we expect a person in their position to do.[9] Our sympathy with their action is in a great measure dependent on a comparison with our habitual standards. Another aspect of this process is that our opinions regarding the behaviour of others are dependent on the context of the actor and his actions. We draw from our experience of society to assign standards of appropriate behaviour; but more than this our own behaviour, and that of others, is shaped by a like socialization. We take on the habitually accepted behaviour of our social positions through practise: and thus we expect a lawyer to act like a lawyer, a clergyman like a clergyman and a soldier like a soldier.

Circumstances habituate us and guide our behaviour just as they guide our standards of appropriate behaviour in others (TMS: 205). Custom, and the habitual conventions which develop with it, are formed in a large part by context and, as we shall see later, this is of central importance to the Scots' notions of social change through time. The Scots' focus on the idea that a practice must be repeated to become habituated leads them to develop a complex theory of socialization to explain the development of customs of behaviour amongst a people. Ideas and practices become general by custom, that is to say that constant repetition leads to conventional expectations and relations developing amongst a people.[10] This, then, is the basis of socialization, the habituation through time, of an individual through interaction with others. Through socialization we come to follow accepted modes of behaviour in a non-deliberative manner which cannot, of itself, be fully broken by the rational reflection of the individual himself, so strongly have our minds become accustomed to it.

From what we have seen it is clear that some part of the social bond is founded on utility, on the notions of gains received by individuals from acting in a social setting. Hume argues that social interdependence compensates for the defects in the powers of each individual (THN: 485), and Smith notes that humans need others to 'improve' (TMS: 13) their position. Social interaction, as well as being 'natural', is beneficial to individuals. But Ferguson is also quick to note that these explanations are not in themselves sufficient to explain the universality of human society. He points out that the social bond is frequently strongest in times of great peril, such as war, when individuals act out of social feeling to defend their group even though in terms of personal utility that action is often unprofitable to them. He argues: 'Men are so far from valuing society on account of its mere external conveniences, that they are commonly most attached where those conveniencies are least frequent' (ECS: 23).

A further principle of sociability is traced by the Scots to human psychology and to the emotional reactions of specific actors. All humans, they argue, are happier in society than in solitude.[11] There is an emotional need for company that is a part of the human psychological make-up. Humans may be able to survive in a desert, and even to prosper there, but they will be miserable until they are admitted to society. This, for Ferguson, is why sociability is part of human nature. The emotional need for society is a deeper explanation for the universality of society than any consideration of utility. Smith traces this emotional need for society to his conception of sympathy. He argues that humans have a psychological need for approval from others. Man 'longs for that relief which nothing can afford him but the entire concord of the affections of the spectators with his own' (TMS: 22). Nor is this desire or need for sympathy to be traced to self-love alone in Smith's view, people do not desire the sympathy of others from considerations of utility or in order to profit by it (TMS: 85–6): they desire it because they need it to function on a psychological level. As Hume noted, the origins of human society are to be found in family groups. This is why, rather than referring to the significance of self-interest, Hume instead prefers the term 'confin'd generosity' (ENQ: 185). That is to say that our concern naturally extends to those close to us, or those related to us.

Moreover, we are more closely interested in or concerned with the interests of others when they relate to our own concerns. Those with whom we live and to whom we are related are closer to our affections and concerns than others more distant from us or unknown to us. Hume defines these circles of concern in the following manner: 'our strongest attention is confin'd to ourselves; our next is extended to our relations and acquaintance; and 'tis only the weakest which reaches to strangers and indifferent persons' (THN: 488).[12] Our exercise of sympathy and concern and interest in the actions and fortunes of others is limited by this confinement of generosity. It is not that we do not sympathize with strangers, but rather that our sympathy with them is restricted by the absence of familiarity.[13] 'Confin'd' generosity is shaped by perspective: our feelings are strongest for those closest to us, those whom we know. Smith's famous example of this confinement is designed to stress the significance of this localized perspective. He writes:

> Let us suppose that the great empire of China, with all its myriads of inhabitants, was suddenly swallowed up by an earthquake, and let us consider how a man of humanity in Europe, who had no sort of connexion with that part of the world, would be affected upon receiving intelligence of this dreadful calamity. He would, I imagine, first of all express very strongly his sorrow for the misfortune of that unhappy people, he would make many melancholy reflections upon the precariousness of human life, and the vanity of all the labours of man ... And when all this fine philosophy was over, when all these humane sentiments had been once fairly expressed, he would pursue his business or his pleasure, take his repose or his diversion, with the same ease and tranquility, as if no such accident had happened. The most frivolous disaster which could befal himself would occasion a more real disturbance. If he was to lose his little finger to-morrow, he would not sleep tonight; but, provided he never saw them, he will snore with the most profound security over the ruin of a hundred millions of his brethren.
>
> (TMS: 136)[14]

We begin to see that the process of sympathy operates by bringing home to us, by rendering closer to our concern, the experiences of others.[15] Self-regarding action is limited by the tendency to sympathize; those sympathetically generated norms of behaviour, which are the product of sociability, act to limit our tendency to follow those inclinations which are the product of our confin'd generosity.[16] As Smith puts it:

> the natural preference which every man has for his own happiness above that of other people, is what no impartial spectator can go along with. Every man is, no doubt, by nature, first and principally recommended to his own care; and as he is fitter to take care of himself than of any other person, it is fit and right that it should be so. Every man, therefore, is much more deeply interested in whatever immediately concerns himself, than in what concerns any other man ... But though the ruin of our neighbour may affect us much less than a small misfortune of our own, we must not ruin him to prevent that small misfortune, nor even to prevent our own ruin.
>
> (TMS: 82–3)

This argument about perspective in moral judgement may be considered in the light of epistemological views on the diffusion of knowledge. The circles of concern are related to knowledge of, and familiarity with those concerned:

our confin'd generosity is a product of familiarity with and concern for those whom we know (TMS: 140, 219). Moreover an individual placed in a particular set of circumstances is, as Smith argued above, in the best position to judge how to act in those circumstances. Subtle moral judgements are best made by those closely related to the circumstances which give rise to them, indeed the whole notion of an impartial spectator is based on knowledge of the circumstances, on 'spectating'.

Smith argues that our natural concern and attention is focused on that which we know and those whom we know. The capacity of the human mind for knowledge and imaginative sympathy is necessarily restricted. As a result of perspective and opportunity we naturally feel strongest towards those familiar to us just as our knowledge of that which is close to us is greatest. While at the same time we find it more difficult to exercise imaginative sympathy with those removed from us, just as we find it difficult to understand knowledge outside our field of experience or specialization. Smith believes that our limited capacities and abilities restrict our attention to a set field of concern (TMS: 237).[17] This, however, will not lead us to a morality of selfishness. As Smith argues the force of sympathy, while attaching us chiefly to those close to us, at the same time limits our actions in regard to others. Human action, whether benevolent or self-interested, is confined in its scope. That is to say that for Smith it is perfectly acceptable to act in the best interests of yourself and those close to you, so long as you do not actively seek to reduce the ability of others to do the same. In this manner those best placed to act in a situation will be responsible for acting to achieve the best outcome; and such actions are reconciled with those of others acting in a similar manner because sympathy and the impartial spectator teach us that it is unacceptable to act directly to harm another.[18]

Society for Smith brings 'tranquillity of mind' (TMS: 23) by providing a guide to our actions determined through the medium of sympathy. Humans, he argues, naturally sympathize with the fortunes and misfortunes of their fellow beings.[19] But as no individual can experience in exact detail the feelings of another, what they do as an act of sympathy is imagine what they would feel in a like situation. Knowledge plays a limiting role on sympathy, we cannot know exactly what another experiences but we can imagine, building up from our own experience and our observation of the situation, what they are going through.[20] However, such imaginative sympathy is necessarily of a lesser degree of emotional strength than that experienced by the person in question. To this end sympathy can only be partial. Humans also, as Smith notes, need or desire the sympathy of others, naturally seeking approval for their feelings and actions. As a consequence individuals limit their emotional responses to bring them closer to that weaker degree experienced by spectators. This is achieved by constructing a mental image of what an impartial spectator would think of our actions and then using this as a guide to what is acceptable or would be approved of by others. By seeking to match the pitch of emotions to that of spectators we develop an equilibrium, or spontaneous order, notion of propriety: a set of conventional or habitual attitudes that guide our actions on a level which will be acceptable to those around us. For this reason Smith refers to society as a 'mirror' (TMS: 110), as a device by which we are able to assess ourselves through others. So the emotional need for sympathy interacts with that propensity for habit formation that we highlighted above, and produces in us a habit of acting in reaction to others' opinions. In this manner our emotional need for approval, and for some reference point for our behaviour in the views of others, leads to a notion of propriety: a notion of how we ought to act and of acceptable behaviour which becomes the basis of praise and blame (and indeed the cornerstone of morality itself).[21] There is, then, a social generation of conventional values which pre-exists written law. Moreover, such habitual conventions become widespread within groups by a process of socialization.[22]

As we have seen, humans are, by the Scots' analysis, highly susceptible to habit, and in particular to the 'habit of society', which Ferguson refers to as a 'habit of the soul' (ECS: 53) and an integral part of what it is to be human. But we have also seen that the interaction of habituated and sympathetic creatures leads to the generation of value systems and conventions of behaviour developed from comparison and mutual adjustment resulting in equilibrium or spontaneous order. Moral value systems are produced intersubjectively or inter-personally. And such customary behaviour affects subsequent actions and our judgements of them. Custom habituates us to certain models of behaviour; indeed the experience of an oft-repeated conventional practice means that 'custom has rendered it habitual to' us (TMS: 201). Such conventional behaviour becomes part of our habitual expectations in the same manner as repeated experience of physical phenomena leads us to form habitual expectations. It is in this manner that 'accepted' or rather expected conventions of human behaviour develop. Born into a society the individual is exposed from childhood to the frequent repetition of attitudes and practices which come to be habitual to them. Such practices, derived from the sympathetic desire to balance sentiments with those expected in the standard of propriety, shape the deliberative education and non-deliberative socialization of children. As Hume puts it:

> In a little time, custom and habit operating on the tender minds of the children, makes them sensible of the advantages, which they may reap from society, as well as fashions them by degrees for it, by rubbing off those rough corners and untoward affections, which prevent their coalition.
>
> (THN: 486)

Such socialization is achieved by repetition of the example and subsequently deepens the custom not only through time in each individual's life, but also through the life of the people in succeeding generations. Hume again:

Whatever it be that forms the manners of one generation, the next must imbibe a deeper tincture of the same dye; men being more susceptible of all impressions during infancy, and retaining these impressions as long as they remain in the world.

(EMPL: 203)

Thus we see again a 'contagion of manners' (EMPL: 204) developing through time and forming the basis of the development of customary behaviour. However, as Smith is keen to note, such socialization pertains within cultures, and as a result the formation of our character within a given cultural or national tradition can affect our judgements of other traditions and cultures (TMS: 195). What this shows us is that context has a vital role to play in the formation of habits and customary behaviour. Humans are socialized within the context of particular circumstances, within a particular society whose attitudes have been in turn formed by the particular circumstances that the people have experienced.[23] A savage is socialized into the circumstances of a savage society, their practices developed in the context of their physical and social surroundings (TMS: 207). And as we have noted, socialization leads to the development of customary expectations, of attitudes and standards of propriety which affect our judgements. Smith offers an example of this in the behaviour which is expected of certain occupations (TMS: 204). The circumstances of a particular occupation shape the conventions of behaviour expected of its practitioners. If they behave in an improprietous manner our judgement of them is affected.

Such concerns of reputation, like the desire for sympathy, lead to conformity of behaviour based around a desire to preserve reputation. In the case of professionals this is not simply to ensure emotional sympathy, but also to ensure business. A laughing undertaker will soon pass out of business as well as become the subject of disapproval.[24] So it appears that utility again plays a role in the development of social conventions. Hume argues that utility and sympathy interact in custom formation; with reference to justice he says: 'Thus self-interest is the original motive to the establishment of justice: but a sympathy with public interest is the source of the moral approbation, which attends that virtue' (THN: 499–500). Moreover, utility plays a role in the process which links our emotional need for sympathy to the circumstances of social life in a manner which moulds the formation of standards of behaviour and practices. That is to say, for a practice to become habituated and socialized by repetition, it must fulfil some use in order for that repetition to occur.

Circumstances

We have seen thus far how great an emphasis the Scots place on notions of habit and custom, especially in the sense of socialization and in their formulation of scientific knowledge. What we must now consider is how it is that customs and habits arise and change. It was noted before that habit is formed by constant experience, by repetition until a phenomenon or practice becomes accepted unthinkingly. But while this explains the process, how a habit is formed, it does not account for why a particular practice becomes habituated into custom. The answer to this question that Hume provides is 'interest'. In other words the practice must have some recurring utility that prompts its repetition. For a practice to pass from a one-off experience into a habituated custom it must be repeated, and for it to be repeated it must successfully fulfil some recurring purpose in relation to its context.

As part of their social science, the Scots set out to examine this process and to discover 'the imperceptible circumstances' (ECS: 65) which lead different groups to have different customs. We have already noted above in our discussion of the Scots' formulation of their 'science of man', that they rejected outright physical determinism in favour of a focus on moral causes. But here we begin to see how the Scots co-opt physical conditions into their analysis. The physical situation in which a people are placed does play a role in the structure of their society, in the formation of their customary and habitual behaviour, but the interesting factor for the Scots is not the direct effect of the physical situation, but rather the indirect effect of human adaptation of behaviour to that environment. Humans adapt their behaviour to their physical situation, and the customs of various peoples are shaped by the challenges that they face. Hume notes that human institutions, such as justice, are the result of an interaction of human nature with 'outward circumstances' (THN: 487). That is to say that the practices which arise in different societies do so as a result of a universal human nature reacting to particular recurring circumstances. Smith follows Hume's argument here and, in *The Theory of Moral Sentiments*, dwells on how habituation to particular circumstances shapes human behaviour. He writes: 'The different situations of different ages and countries are apt . . . to give different characters to the generality of those who live in them' (TMS: 204).

Thus, for Smith, a savage becomes inured to hardship and their behaviour becomes shaped by their circumstances. As he puts it: 'His circumstances not only habituate him to every sort of distress, but teach him to give way to none of the passions which that distress is apt to excite' (TMS: 205). So experience teaches a savage the most profitable way to act, they are socialized by the example of others who have similarly learned from experience those practices necessary for survival. This behaviour is not unique to savage states: Smith also applies the analysis to commercial societies through the example mentioned above on the behaviour of different occupations. Thus customary modes of behaviour become associated with certain professions and this behaviour is determined by the circumstances of the profession, the role played in it by the individual and the response to that role by others. The behaviour we come to associate with clergymen and soldiers is different, as the circumstances of their professions and their socialization into the customs of that profession differ. Through sympathy a concept of proper behaviour is formed and a notion of propriety developed according to the circumstances of each occupation. This then becomes part of the individual's professional reputation upon which they trade for their livelihood. As we noted above we would not expect an undertaker who laughed constantly to be much of a financial success, indeed we would heartily disapprove of their improprietous behaviour, but we would not think the same of a jovial publican. Here we have an example of propriety acting as an invisible hand to produce a benign spontaneous order.

Adaptation to external circumstances accounts for much of the diversity to be found among peoples. 'In consequence of habit,' Ferguson writes of humans, 'he becomes reconciled to very different scenes' (ECS: 200).[25] But the circumstances in which humans find themselves are not solely physical in their nature. They are also social (or moral). Humans exist, as we have seen, in a social context, and experience those conventions of behaviour that have been formed by their predecessors and contemporaries. Such conventions of behaviour as are already existent have been formed in relation to circumstances, or 'accidents' as Ferguson (ECS: 123) calls them, and have been repeated because they have been found useful. Individuals become socialized into a culture and habitually accept these conventions. This process though does not imply either an explicit agreement with or endorsement of these practices by each individual.[26] We are dealing with habitual acceptance of circumstances, thus the habits of others become part of the circumstances to which we become habituated as we are socialized. We need not have any conscious notion of the utility of these practices, but our propensity to form habits and desire for social acceptance lead us to accept them without any great thought. In Hume's example of the two men in a boat we see how human behaviour regulates itself and comes into co-ordination to secure a useful end without conscious, rational or explicit agreements taking place. The circumstances – two men in a boat – and the end – to cross the body of water – shape the behaviour of the actors and, when often repeated, lead to a convention becoming habitual.[27]

This focus on the importance of individual situation is a prominent feature of Smith's *Theory of Moral Sentiments*. The passages on stoicism in Smith's work stress an approval for the adaptation by individuals to the concrete reality of their situations. He writes: 'The never-failing certainty with which all men, sooner or later, accommodate themselves to whatever becomes their permanent situation, may, perhaps, induce us to think that the Stoics were, at least, thus far very nearly in the right' (TMS: 149). What develops from this is an analysis of the virtue of prudence (TMS: 262). Ferguson (1973 vol. 1: 232) defines prudence as the habit of adapting to circumstances, and Smith lays it in a particular relief, arguing:

> The man who lives within his income, is naturally contented with his situation, which, by continual, though small accumulations, is growing better and better every day ... He confines himself, as much as his duty

will permit, to his own affairs, and has no taste for that foolish importance which many people wish to derive from appearing to have some
influence in the management of those of others.

(TMS: 215)[28]

Self-command becomes a virtue as it teaches humans to adapt to the specifics
of their particular local situation, and though prudence is a 'selfish' virtue, it
is not in any sense detrimental to society. It is simply a matter of restricting
attention to that epistemological field which is within the grasp of a particular mind.[29] If you like, it is a form of specialization. Our knowledge of our
own situation is necessarily greater than that of any other person, and so
each individual is best suited to utilize their specialized knowledge to make
effective decisions. Moreover, prudence links into the natural 'confin'd generosity' of individuals and allows them to act efficiently within the scope of
their most intimate passionate concerns. In a commercial society the
prudent individual acts within their means: frugality and temperance
become virtues as they aid this restriction of behaviour to such a field as may
effectively be influenced by each individual given their particular circumstances (TMS: 28).

   As we have seen above, Smith argues for a comparative, inter-personal,
development of moral standards through the medium of an impartial spectator: the desire for praise and praiseworthiness shape man's understanding of
acceptable behaviour in a social context. The standards that develop are
those of propriety – that which is acceptable social behaviour is determined
by consultation with the impartial spectator. It is in this manner that
humans restrict their emotional displays to those acceptable to those around
them, limiting the 'pitch' of their emotions to that which is suitable to
engage the sympathy of their fellows. Habit and experience teach individuals the standards of propriety which exist in a given society, in this
sense the psychological need to please, itself a facet of human nature, renders
a sense of propriety 'natural' to humans. Prudence is a virtue that arises from
humans adapting to their particular circumstances and acting in an efficient
'economic' manner: while propriety is a virtue which arises from humans
adapting to their particular circumstances to act in a socially acceptable
manner and both of these form a part of the invisible hand that promotes
socially benign spontaneous orders. In this sense the natural sympathy that
Smith describes is limited by the circumstances of the individual, with these
circumstances rendering extremes of sympathy improprietous given the
socially generated standards of behaviour (TMS: 140). Prudence and propriety have to do with individual perspective: sympathy and the impartial spectator being the psychological media that allow humans to extend their
concern beyond their immediate circle while maintaining an appropriate
and socially acceptable degree of perspective and detachment.

   One interesting result of this approach to the generation of moral values
is that it leads to a focus on 'moderate virtue' (Clark 1992: 187); or as

Mizuta puts it: 'It is not so much the excellent virtues as ordinary propriety that Smith is trying to explain as the main subject of his book' (Mizuta 1975: 119). Instead of pursuing an approach that seeks to define and justify the nature of virtue, Smith set out to explain the actual generation of the modes of behaviour that facilitate social interaction. Clark argues that this was a result not solely of the explanatory approach of the Scots, but also of their particular stress on the conventional, spontaneous order nature of the generation of customs: 'Smith attributed to commercial society a kind of moderate virtue, less dazzling than that of the saint, the sage or the state-builder, but more useful, because more frequently activated' (Clark 1993: 345). The process of 'adjustment and compromise' (Clark 1992: 202) in search of stability leads to a high degree of conformity in matters of every-day interaction. The submission to such interpersonally generated forms of behaviour eases interaction and reduces uncertainty.[30]

Custom becomes a part of the external circumstances to which we must adapt and socialization makes this experience relatively simple for the bulk of mankind. As Hume and Smith stress, habitual behaviour becomes ingrained and hard to shift. Customary behaviour, insofar as it shapes part of the circumstances in which we find ourselves, becomes hard to change. Humans continue to act in a habitual fashion even after the circumstances from which that habit arose have changed (ECS: 132; WN: 380). There is a problem here. A custom is a habituated practice drawn from experience whose end is utility based on a certain set of circumstances. But the strength of habit and custom, added to by long practice and socialization, is such on the human character that even after those circumstances shift the behaviour pattern lingers on. How, then, does an individual accustomed to a savage state progress to civilization?

We can conjecturally reconstruct Smith's answer to this question by examining his discussion of the practice of infanticide. Smith, Hume and Ferguson all make reference to this ancient practice as an example of how a morally reprehensible (in their view) practice can become accepted by even relatively advanced people.[31] The explanation that they have for the origins of this practice is once again grounded in utility. It is a response to popu-lation growth in a situation of limited physical resources (EMPL: 398; ECS: 135). This form of behaviour, which Smith believes is contrary to human nature and feeling (TMS: 210), became habitually accepted: humans put aside their horror in reaction to their circumstances, and the repetition of this practice rendered it a custom which became accepted by the people as a whole. As Smith notes, such behaviour is more understandable, or prudent, in situations of extreme indigence where the survival of the parent is also at stake. However, once the circumstances of subsistence change the practice continues as it has become ingrained. Smith argues:

> In the latter ages of Greece, however, the same thing was permitted from views of remote interest or convenience, which could by no means

excuse it. Uninterrupted custom had by this time so thoroughly autho-
rised the practice, that not only the loose maxims of the world tolerated
this barbarous prerogative, but even the doctrine of philosophers, which
ought to have been more just and accurate, was led away by the established custom.

(TMS: 210)[32]

Though Smith condemns this behaviour he recognizes that it cannot long
survive, arguing that no society can persist in customary practices which go
against the tenor of human sentiment and feeling (TMS: 211). How, then,
did this custom pass out of use, how did people free themselves from socialized acceptance of the custom of infanticide? Smith's theory suggests that
people gradually became aware of the incongruity of infanticide with human
nature and feeling, they came to be repelled by it, and as their material condition became more secure, rejected it as a practice to deal with issues of
population. The customary practice failed because although it supplied an
answer to the circumstances based on utility, it failed to balance sympathy
against utility in answering the problem when those circumstances had
altered.

Such a process relies, on at least some level, on a deliberative calculation
which balances sentiment with circumstance. What is significant for Smith's
theory though is that such deliberative judgements occur in relation to
particular cases of infanticide. They do not refer to a process of the rational
examination of the practice as a social phenomenon, but rather to a feeling of
repulsion leading to a belief that the practice is unacceptable. The moral
philosopher and politician are of little importance in this account of moral
change.[33] Instead Smith's model deploys an individualistic micro-level
explanation to account for the macro-level change in social practices. Individual actors make deliberative decisions to cease the practice of infanticide
based on their examination of their changed economic position and the
incongruity of infanticide with the universal human emotional attachment
to children. The specific individuals who cease to practice infanticide then
internalize this stance through the impartial spectator and begin to form the
opinion that the spectator would not approve of infanticide as either prudent
or fitting with propriety. This internalization through conscience leads these
individuals to judge the behaviour of others within the group as unacceptable when they practice infanticide. Gradually, as more individuals become
aware of the incongruity, those who continue the practice begin to feel the
disapprobation of increasing numbers of their fellows and begin to moderate
their behaviour in order to secure the approbation of their fellows.

It is important to note that the motivating factor in the assessment here
is not utility but rather emotion and the desire for the approbation of our
fellows. The internalization of the new standard of propriety as an aspect of
individual conscience results in a situation where individuals limit their
actions before the deed to avoid disapprobation. Over a period of time the

practice gradually falls from use and there is a gradual shift in the conventional behaviour of the group. The new convention becomes habitually accepted, generalized and forms the basis of individual assessments of proper behaviour. Individuals become socialized to accept the new standard and there is a gradual shift in moral attitudes.

The important thing to draw from this is that custom and habit do indeed retain their force after circumstances change, but only for so long. The initial custom must have had some grounding in utility and should external circumstances change to remove the utility of a practice, it will not be long before people evolve new customs in reaction to the new circumstances. As a result they will modify their behaviour to the new circumstances (adjust the equilibrium), and adapt those practices to form new customs. This slow and gradual process is what the Scots regarded as the progress of manners, a phenomenon which represented a key indicator of the progress of civilization.

It is clear that the Scots adopt a spontaneous order approach to the explanation of the origins of morality. Drawing on the underlying universality of human sociability and the desire for stability of expectations, they identify a process of sympathetic mutual adjustment in reaction to circumstances (prudence) and to the views of others (propriety). Our moral rules are a spontaneous order that arises as an unintended consequence of our habituation to such modes of behaviour. The order itself evolves in a gradual manner in reaction to changes in the circumstances of the people in question. Throughout this approach to the morality there is a rejection of purposive rationality and an assumption of the significance of the limits of the human capacity for knowledge. Moreover, this analysis of the development and operation of morality, with its stress on the emergence of social stability, forms a part of the invisible hand argument that explains the generation of socially benign spontaneous orders.

4 The science of jurisprudence

The four stages

Having developed a position that acknowledges the centrality of habits and conventions formed in reaction to circumstances, the Scots proceed to an analysis of the spontaneous emergence of key social institutions such as property, law and government, which is conducted in the light of these insights. They deploy the spontaneous order approach through a conjectural history aimed at the explanation of these institutions.

For the Scots, and particularly for Hume and Smith, notions of property and justice are coeval. Their origins are intimately related and indeed explain each other. For Hume the notion of justice can only exist where the conditions for it exist, and as these conditions are the same as those which produce conventions of property, then the origins of the two concepts are mutually explanatory (THN: 491).[1] The Scots' analysis is grounded on the premise that human beings form conventional modes of behaviour in reaction to their circumstances. Perhaps naturally the chief focus of human attention is the provision of subsistence. Humans require sustenance and shelter to survive, and as a result these matters become the focus of a great deal of their actions. As Ferguson notes: 'the care of subsistence is the principal spring of human actions' (ECS: 35), food is a product of human industry, that is to say humans must act in some way to secure it for their consumption. Thus in all societies the provision of subsistence is the 'prior' industry (WN: 377), for without it the survival of the species is impossible and thus other activities are equally impossible. In undeveloped, or savage, nations the individual's first concern is survival. As a result their first efforts are to secure subsistence. This, in the Scots' view, accounts for the immediacy of savage societies. The difficulty of securing subsistence leads humans to focus their attention almost solely upon it. The conventions and forms of human behaviour are shaped in great measure by the various devices which they develop to provide for their subsistence. This, the Scots argue, is a universal phenomenon, occurring as it does in all human societies. The underlying universality of the human need for subsistence, combined with the similarity of our physical frames, nature and intellect means that the development of different modes of subsistence is a process that occurs in a similar manner in all human societies. As Millar puts it: 'the similarity of his wants, as well as of the faculties by which those wants are supplied, has everywhere produced a remarkable uniformity in the several steps of his progression' (Millar 1990: 3). From this insight, supported and confirmed by the evidence of conjectural history, the Scots develop their stadial theories of progress and social change.

The most clearly defined stadial analysis is that developed by Smith, and mirrored by Millar, which has become known as the 'four stages' theory.[2] Smith divides types of society into four categories based on their reactions to the issue of subsistence. The mode of subsistence, he argues, as the primary concern of human activity, necessarily shapes other social institutions that develop in each of these types of society. Smith's four stages; '1st, the Age of Hunters; 2ndly, the Age of Shepherds; 3rdly, the Age of Agriculture; and 4thly, the Age of Commerce' (LJP: 14) are laid down as a general schema of social development which is applicable to all societies. Thus each stage produces conventional, including as we have seen 'moral', behaviours that are appropriate for the physical conditions and level of security of subsistence which pertain in them. It is through this conceptual framework that Smith and Millar approach the gradual development of social institutions. Millar's *Distinction of Ranks* is a series of case studies of basic human interrelations and how each is affected by the gradual change in the mode of subsistence. He examines notions of subordination, the position of women, the relation between parent and child and that between master and servant, tracing in each case the changes in the conventions and attitudes around each relationship in the light of changes in the mode of subsistence.[3] In each stage a different method for securing subsistence dominates: hunting, herding, agriculture and commercial industry. But each stage also absorbs the stage before: hunting and herding do not cease because agriculture arises, but they cease to be the sole or chief means of securing subsistence. For this reason Smith argues that in a commercial society hunting and fishing persist, but as non-essential activities undertaken for pleasure rather than through necessity (WN: 117). By examining the evidence of conjectural history the Scots determined that all societies, if left alone to develop, proceeded roughly according to this pattern of change in the mode of subsistence.[4] Change between each of the stages is posited on the discovery of new skills which prove more productive in securing subsistence than those developed in the past.

Smith argues that animals multiply in direct proportion to the provision of 'subsistence' (WN: 97): as a result there exists a constant demand for food owing to universal, natural, drives for procreation and survival. As subsistence becomes more secure in each stage the population grows as larger families may be supported (WN: 98). However, population growth itself cannot be the reason behind a change in the stage of the mode of subsistence.[5] It certainly may act as a prompt to that change, but the means depend on the acquisition of the knowledge requisite to pursue the new mode. Thus the change from hunter to shepherd is brought about by the gradual development of the skills necessary for animal husbandry: similarly agriculture is a skill acquired in reaction to circumstances. That is to say that the desire to supply more steadily the means of subsistence for a greater population led to experiments in food production which led in turn to the discovery and refinement of new methods. Millar notes this when he attributes the changes between stages to 'experience' (Millar 1990: 3). Ferguson in turn argues that the development of new skills in the provision of subsistence is related to the gradual rise of settled communities.[6] Thus a shepherd society gradually settles in one geographic location, their growing familiarity with that situation opening their attention to possible new means of subsistence and leading to the honing of new skills in agriculture. If we are to tell the story of the links between the 'four stages' in terms of the rise of knowledge of means of subsistence it would be something like this: Hunters are brought into repeated contact with animals and gradually acquire the skills which form the basis of shepherdry; shepherds are brought into contact with the means of subsistence of animals and gradually acquire knowledge of the crops required, their attention is then led to a possible new source of human subsistence and, as they settle geographically, they develop agricultural skills. Once humans have developed settled accommodation the division of labour increases and commercial industry begins to develop.[7]

Smith is also aware of this when he compares colonists to savages: 'The colonists carry out with them a knowledge of agriculture and of other useful arts, superior to what can grow up of its own accord in the course of many centuries among savage and barbarous nations' (WN: 564–5). Thus we see that the key factor here is the possession of knowledge. Colonists draw on the experiential knowledge of their mother country to provide for their subsistence. Savages on the other hand have yet to pass through the 'four stages' and acquire the gradual development of knowledge relating to subsistence that it entails. Smith states that the population of a country is a mark of its prosperity (WN: 87–8): that its development of subsistence provision allowing it to support an increasingly large population is the measure of both its wealth and 'progression'. Why this should be so is a point to which we will return later in the next chapter: but here suffice it to say that for Smith the extent of the division of labour is limited by the extent of the market which is, in turn, determined by population. The division of labour allows the growth of specialization, specialist knowledge and efficiency which, when related to the primary object of subsistence, increases the scope for further growth of population.

This analysis of social change and of the effect of the mode of subsistence on the nature of society and population is the backdrop to the Scots' discussion of the interactive development of property, justice and government. The Scots define property as a mental concept, as something which 'is not anything real in the objects, but is the offspring of the sentiments' (THN:

509). Property is not a physical relationship, but an artificial convention of human behaviour that develops in relation to a specific set of circumstances and to serve a specific function.[8] Property is 'such a relation betwixt a person and an object as permits him, but forbids any other, the free use and possession of it, without violating the laws of justice and moral equity' (THN: 310). In terms of the 'four stages' theory it is clear that no abstract conception of property exists in the savage or hunter society. The immediacy of life in such conditions precludes abstract thought and the mode of subsistence is based around securing from what is wild for immediate needs. As a result of this there is little or no government in hunting societies. As Smith puts it 'Till there be property there can be no government' (LJP: 404). When there is no mode of distinction between a people, no dependence or security of subsistence, then the concept of property and of a government to defend that property is absent.

However, government and property do arise, and they do so in the second of Smith's stages, that of shepherds. It is in the age of shepherds that notions of property, subordination and government arise and the development of social institutions truly begins. Experience teaches individuals to refine the skills necessary for animal husbandry as they see the benefit in this domestication of animals as opposed to constant hunting. But shepherdry is discovered and perfected by some before others. These people control increasing numbers of animals rendering hunting increasingly difficult for others (LJP: 202). However, this situation was as yet insecure. The shepherd might domesticate and tend his flock, but his claim to them as a result of this could quite easily pass unnoticed by others keen to secure subsistence (LJP: 404). Some institution to enforce claims of right was required by shepherds. The origin of that institution was also to be found in this inequality of fortune, for those who could not practise shepherdry and yet saw the stock of wild animals fall would become dependent on those who had mastered the skill. Those who controlled herds and flocks came to occupy superior positions as an unintended consequence of their possession of the knowledge of shepherdry; knowledge which gave them easier access to subsistence through the control of large numbers of animals. However, the control of large numbers of animals is in itself useless because of the physical limits as to how much each individual can consume. As a result the successful shepherd provides for others who have yet to acquire the skill and, consequently, comes to a position of eminence over them and introduces subordination into society for the first time. Dependants develop a habit of obedience and accept their position as clients in order to secure easy access to the means of subsistence. They come, as a result of this process of habit, to accept the validity of the shepherd's claim to his flocks (LJP: 405), forming an opinion of his 'right' to the control of them. They also begin to develop an emotional loyalty to their particular benefactor and his heirs (WN: 715) that is the foundation of a notion of a nation, or the explicit identification with institutions which express the unity of the community. The first institutions of government arise with the explicit purpose of defending property and are supported by the dependency-led obedience of people to those who have acquired flocks and herds. In the second stage wealth supplies authority (WN: 713) and introduces an inequality into human society which bears little relation to physical attributes.

It is in this manner that the 'habit' (ECS: 81) of property arises and is adapted in each of the succeeding stages, gradually being refined to deal with the particular circumstances of each new mode of subsistence and the events that occur in the course of its development. Property is a spontaneous order, a settled equilibrium that gradually evolves in reaction to changes in circumstance. Thus, Smith argues, in the age of shepherds the centrality of the concept of ownership leads to property laws which punish theft with death. This practice passes out of use in the age of agriculture where theft is no longer such a direct threat to subsistence. In the age of commerce the huge increase in the scope of those things which can be held as property leads to a proportionate increase in the laws to defend that property, though once again theft is no longer considered of such an immediate threat to subsistence as to give occasion for the death penalty (LJP: 16).

In the age of shepherds the conception of property refers to herds and flocks and the wandering nature of such peoples precludes any definite notion of property in land. But in the age of agriculture property in land develops in reaction to the fixed habitation of agricultural labourers. For Smith the key step in the development of private property is the development of fixed habitations in cities and towns (LJP: 408–9, 460). This phenomenon is merely a continuation of the concentration of population that had proceeded from hunting through shepherding ages. In a hunter society social groups are relatively small, each competing for the scarce resources of the hunt. Shepherd societies admit of larger numbers by the greater ease of subsistence, but these numbers do not settle in a specific location to practise their arts. They are, however, open to attack by other groups and so, for reasons of mutual defence, erect fortified towns to which they might take their flocks to avoid attack (LJP: 409). The concentration of population in these locales leads to a development of the arts, in particular agricultural skills; and towns and cities come to develop. As a result of this: 'Private property in land never begins till a division be made from common agreement, which is generally when cities begin to be built, as every one would choose that his house, which is a permanent object, should be entirely his own' (LJP: 460). The concerted development of private property is to be found in early urban areas where people living close together found it necessary to define their separate possessions. The notion of private property arose gradually from a sense of the utility of the mutual recognition of claims of right to property.

The clearest description of the details of this process in the work of the Scots is that given by Hume.[9] Though Hume does not explicitly follow the 'four stages' schema, he does nonetheless provide a conceptual analysis of how the concept of property arises from utility. Smith was keen to stress that the stability of property is necessary for the stability of society (LJP: 35), and Hume's analysis provides a theoretical explanation of why this is so. We have already seen that Hume shares the common Scots' view that the origins of property and justice are interrelated. He believed that justice does not arise from a sense of benevolence – this having too weak a hold on the human imagination – but rather that it is a product of individuals' self-interest. We desire to acquire goods to support our subsistence and indulge our appetites, but the danger of this urge is that it makes society impossible. That is to say if we do not refrain from others' goods, if we constantly seize them, we are left permanently open to the threat that the same will happen to us. To this end humanity discerns the utility, in the long term, of refraining from the goods of others. As Hume puts it:

> Now this alteration must necessarily take place upon the least reflection; since 'tis evident, that the passion is much better satisfy'd by its restraint, than by its liberty, and that by preserving society, we make much greater advances in the acquiring possessions, than by running into the solitary and forlorn condition, which must follow upon violence and an universal licence.
>
> (THN: 492)

What arises though is not an explicit agreement over the security of property, but a convention of behaviour. By experience we gradually come to see the value of such an institution to promote stability and permit the growth of wealth.[10] Like a habit, the convention: 'arises gradually, and acquires force by a slow progression, and by our repeated experience of the inconveniences of transgressing it' (THN: 490). The gradual increase in experiential knowledge leads humans to form customs the utility of which are the security of possession and the consequent scope for future profit. Our expectations become stabilized, we rely upon private property conventions to reduce the uncertainty of our position. For example, if I know that my neighbour will refrain from stealing my goods so long as I do the same toward him, then I need not stay up all night on guard.

However, there is a problem with this model, as it is laid down thus far, and it is this problem which leads to the development and evolution of enforceable rules of justice and the institution of government. Hume argues that all human societies are subject to the same conditions which render justice necessary: these being 'the selfishness and confin'd generosity of men', 'the scanty provision nature has made for his wants', and the 'easy change' of external objects (THN: 494–5). These universal conditions mean that all societies will eventually develop some sense of property once the utility of such a conception becomes apparent. The easy interchange of goods together with limited generosity and the scarcity of those goods create a problem for human interaction which they seek to solve by the institution of private property. Justice is an artificial virtue that arises around this convention; abstinence from the property of others produces long-term utility and this is the original authority of justice. Such conventions become part of the customary behaviour of a people by repetition and socialization, and they gradually come to affect the sentiments of the people. As we have seen, individuals seek the approval of others in terms of sympathy, and so conform to those social norms and conventions within which they have been socialized. Thus justice, whose original motive is utility, becomes a virtue (THN: 499–500). Justice and property are both artificial, yet they are both vital to the continuation of society.[11]

The origins of government

Conventions develop regarding property and Hume is quick to assert that these rules must be general in nature if they are to be accepted as conventions by all. This focus on the need for general rules is yet another product of the human propensity to seek systematized knowledge, classification and order, to stabilize expectations. Such general rules, as with habitual expectation, create a conception of probability drawn from experience. Or as Hume puts it: 'General Rules create a species of probability, which sometimes influences the judgement, and always the imagination' (THN: 585). Humans are 'mightily addicted to general rules' (THN: 551), and this emotional need for order and systematized knowledge leads them to form general rules drawn from their experience. General rules of behaviour are evolved from experience of circumstances, but it is difficult to produce rules with a specific content which apply in more than one particular set of circumstances. For this reason humans categorize and generalize, they develop abstract rules which are not content or context specific, and as a result may be applied to a variety of circumstances.[12] As Hume would have it:

> All the laws of nature, which regulate property, as well as all civil laws, are general, and regard alone some essential circumstances of the case, without taking into consideration the characters, situations, and connexions of the person concerned, or any particular consequences which may result from the determination of these laws in any particular case which offers.
>
> (ENQ: 305)

General rules can be distilled from complex situations by the observation of broad regularities. In a sense, simple general rules, insofar as they are attempts to create a stability of expectation, are a reaction to complexity. Our 'laws of nature', as with any law, are general rules drawn from the observation of experience, or acquired through socialization with others acting on their experience. They create a sense of certainty and stability of expectation that calms the mind. For this reason conventions such as property develop as general rules to stabilize interaction. The stability provided by a system of general rules in society is, as a result, more important to the success of that society than the outcome of specific individual cases where the rule is applied (ENQ: 304). It is because of this, and because agreement to the institution of property laws is necessary, that the laws governing property take on a general form.

The rules of property, in order to stabilize expectations, must be such that ownership can be determined. The Scots lay down various criteria that develop as conventional claims to property: first possession, long possession, inheritance and alienation (in the sense of buying and selling). From this the Scots believed that disputes over the validity of claims made to property by appeal to these principles would naturally arise (LJP: 203). As a result societies would have to develop some conflict resolution mechanism if such disputes were not to tear the society apart. Also, external goods remain interchangeable by their nature and this factor, together with the other conditions of justice, means that there is a constant temptation to break the convention for private and immediate gain. Some institution must be developed to secure the stability of property that is necessary for the stability and indeed the existence of society.

The fact, arising from the conditions which form the conventions of justice and property, that society requires some support for the non-physical claim which is the convention of property implies that the first law is law determining and governing property (LJP: 313). This law exists as convention and custom, but – just as we saw the utility of property – so we see the utility of a body to determine and decide in disputes regarding property. The need to delimit property in an accurate manner, in order to avoid potential conflicts that would destroy society, leads to the institution of government (ENQ: 192). As the conditions which render justice necessary and give rise to the 'habit' (ENQ: 203) of property are universal factors of the human condition, we see that all societies institute a conception of property and develop conventions which decide property disputes under the name of justice.

Therefore the institution of government arises from the recognition of a common court of appeal for the settlement of property disputes.[13] The conventions of property ownership that arise in a society thus begin to be codified, to become laws, when they are drawn up and made explicit by those appealed to as judges in disputes. This process, the desire for general rules and stability of possession, though it is prompted by a sense of interest arising from a view to utility, is not carried on in any explicit and intended manner.[14] Those who appeal to a judge to decide disagreements over the conventional rules of property do not intend to create the institution of government. Just as those who undertake specific moral judgements do not intend to create a system of common morality. As Ferguson puts it:

> Mankind, in following the present sense of their minds, in striving to remove inconveniences, or to gain apparent and contiguous advantages,

arrive at ends which even their imagination could not anticipate, and pass on, like other animals, in the track of their nature, without perceiving its end. He who first said 'I will appropriate this field: I will leave it to my heirs' did not perceive that he was laying the foundation of civil laws and political establishments.

(ECS: 119)

The Scots believed that governments arose by a complex process of unintended consequences in reaction to circumstances, the chief of which being the desire for property delineation to secure subsistence and the desire for protection from external threats. Government, however, develops more slowly than the arts of subsistence (WN: 565), its attentions being called upon only when disputes arise. In the meantime the advance of knowledge of the arts of subsistence grows. Smith is particularly clear that the chief scene of the advance of knowledge of both government and the arts is in the burgeoning cities where interaction and trade develop productive techniques and institutions (WN: 405, 411). People living in close proximity have more scope for conflict as well as for trade. Thus government develops to more advanced levels in urban areas. However, cities require to trade their produce for that of the country in order to acquire some of the means of subsistence; for this reason Smith spends some time analysing the relationship between town and country. Local farmers can come to town to trade on market day, but as trade between communities advances immediate exchange becomes unwieldy and the notions of contract and money arise (LJP: 91). The enforcement of contracts within a given area then becomes the rationale behind the extension of the judicial power of governments. Moreover, the need for stability and peace to allow the advance of learning in the commercial arts means that commerce gradually alters the practice of government reducing the scope for arbitrary uses of authority (WN: 412).

Hume's analysis of the origins of government and property in the *Treatise* and *Essays* describes the underlying rationale for such institutions; it explains how interest and a sense of utility are the original spurs to humans establishing government and property. But the actual development of these institutions is laid out in terms of the formation of conventions. And these conventions arise by a series of particular reactions to historical circumstances that become habitual. There is no purposive creation of government, no pre-recognition of the utility, the sense of its usefulness arises as it is practised and developed. It becomes habitually accepted because it is repeated and because these repetitions stabilize expectations. Hume highlights this when he argues that though humans are sensible to the long-term advantages of a system of justice governing property and applied by an institution of government, they are also by their nature weak. They are predisposed to view matters in terms of short-term advantage (THN: 534–9). It is because of this that an individual can understand the utility of justice yet act in a manner pernicious to it. This 'narrowness of the soul, which makes them prefer the present to the remote' (THN: 537) is a facet of human nature which leads to the constant threat that those conventions of justice adopted out of a sense of utility will fail to bind. There must, then, be some principle that binds a society to justice, which overrides this short-termism and encourages the view of longer-term utility. This human partiality towards our short-term goals is related to the evolution of property in terms of general rules. Hume argues that general rules of behaviour restrict the operation of our short-term self-interested urges and force us to act in a manner that keeps the long-term advantages of institutions such as property in view (THN: 597). What Smith calls 'general rules of conduct' (TMS: 161) restrict our capacity to act in a specific manner in reaction to specific circumstances. By circumscribing ourselves with general rules we are able to keep longer-term utility at least partly in view.

The Scots' analysis focuses on the notions of habit and convention, in the habit of obedience or the acceptance of authority. This analysis, while aware of the function of utility in the underlying rationale of the process, highlights how the actual development occurred through a process of evolution from the unintended consequences of the human desire for order and stability of expectations. The desire for adjudication of disputes about property leads people to turn to eminent individuals within their society, individuals whose fame – though not originally based on impartial judgement – suggests them to the imagination as judges. In his essay on the *Origin of Government* Hume describes how this process occurs in some detail. He argues that: 'The persons, who first attain this distinction by the consent, tacit or express, of the people, must be endowed with superior personal qualities ... which command respect and confidence' (EMPL: 39). We see that the origin of government, though grounded in a rationale of utility, in fact develops in a gradual manner: it evolves from the habitual acceptance of chiefs.[15] The role of the chief is an evolved institution grounded on habitual acceptance, it is not the product of a rational plan or contract grounded on an explicit attempt to secure a useful end. Hume makes this point clearly: 'it cannot be expected that men should beforehand be able to discover them, or foresee their operation. Government commences more casually and more imperfectly' (EMPL: 39).

Though the rationale behind motivations of the self-interested actors who appeal to a common judge is the desire for property delineation, the process, when repeated, creates a new authority in society, it introduces a 'casual subordination' (ECS: 129) which grows into a natural deference to the decisions of the chief. Such is the force of habit among humans that they come to form the view that 'Antiquity always begets the opinion of right' (EMPL: 33). Through the force of habit people come to recognize the authority of a chief or government even though the origin of the particular chief or government's power would undoubtedly have had little to do with the utility of stable property.[16] Likewise, in Hume's *First Principles of Government*, the origins and justification of government are a complex interrelation of factors that coalesce around individuals of ability who are in the right place at the right time, and then develop through time into the modern institution of government. Ferguson also notes that this process of habituated subordination and opinion of right through time is a factor that is socialized in each member of society. The 'contagion of society' (ECS: 156) creates social bonds among a people which combine with their moral approval of the claim of being a people. Thus each people begin to view the form of government as an intrinsic part of the nation, a symbol of what they as a people have achieved. To this extent the origins of that government are not important, except in so far as they display the achievement of a particular people.

Government is a convention that arises within a society, is authorized by the utility of a central authority to determine property and is strengthened by the force of habit and custom. The institutions of justice need not be just in their origins. Indeed the story, as the Scots have told it, of the origins of government clearly abandons the focus on individual purposive action that marked those simple models of explanation that they had already rejected. The instigation of government was a product of a process of unintended consequences: it arose from the temporary reactions to present concerns and literally 'grew' from there in reaction to new circumstances. As time passed these conventions among humans became habitual and possessed a force in their minds which, though the institutions themselves were ultimately underwritten by notions of utility, carried an emotional strength which made them part of the social bond. The institution of government is shaped by the human reaction to the circumstances in which mankind finds themselves. When people come to live in close proximity they begin to interact in terms of private property relations and, government arising, material progress proceeds. Law arises from the decisions of 'judges' in particular cases and a convention of following precedent arises that has the effect of stabilizing expectations and allowing social interaction to occur with minimal levels of uncertainty.[17] The institutions of government and law are a spontaneous order, they arise from local conventions and serve to place order on everyday life, and are accepted as such for that reason. Justice, as with science, springs from a desire for systematized knowledge, a desire to reduce uncertainty. So too does law fulfil this function of calming the mind, of leading our habitual thought processes in an ordered manner in line with our expectations. Laws are adapted to the particular circumstances experienced by a people, and they are determined by these circumstances and humanity's reaction to them. As we have seen, the need for justice is universal owing to the universality of the conditions of justice, but the specific reaction to these universal phenomena is dependent on the particular conditions of each society. It is because of this that forms of government and law differ. Indeed much of Smith's *Lectures on Jurisprudence* is taken up with a comparison of the laws of differing nations, showing how the particular circumstances moulded their reactions to common problems. But Smith also applies his four stages of analysis to this, in particular showing how each law develops in line with commerce and how commerce encourages stable government as much as it requires it (WN: 412). Government and law have a progress, just as the mode of subsistence has, and the Scots devoted considerable attention to this notion of progress.

In terms of historical change, human institutions develop in reaction to changed circumstances, and the diversity of human institutions increases with their increased complexity. Indeed the character of social institutions and human character more generally are formed by the experiences acquired in different circumstances. It would appear that such reactions to circumstances by social groups lead to a situation where: 'The multiplicity of forms ... which different societies offer to our view, is almost infinite' (ECS: 65).

Despite this diversity of appearances the Scots believed that a 'science of man' was possible: that a general theory of society could be developed from the significant similarities between societies in similar stages of development. As we saw above, the stadial theory of social development and the Scots' analysis of the origins of government and property were conducted in the light of what they saw as universal factors that underlay this diversity. Human nature and the conditions of justice could, in their view, be reduced to a few general principles that applied to humans in all their circumstances. Diversity is constituted by the means (both institutional and conventional) of dealing with these universal factors within a specific context. Or as Hume puts it, speaking of human institutions: 'which cause such a diversity, and at the same time maintain such a uniformity in human life' (THN: 402). The Scots' stadial theories and conjectural history are underpinned by a series of universal principles of nature and human nature (themselves identifiable by the comparative method and conjectural history) that can be developed upon to provide a 'scientific' basis through which to approach society. Context does indeed shape human experience, but underlying universals such as the desire for subsistence and the conditions of justice lie behind every circumstance in which mankind is found. Where the differences arise is in different social groups' reactions to them.

Progress

In addition to being a theory of social change the Scots' stadial approach is also a theory of progress. From similar barbaric origins some human societies 'improved', became 'polished' (TMS: 208) and civilized as a result of their having passed through stages of development typified by the mode of subsistence. The Scots believed that man has 'a disposition and capacity for improving his condition, by the exertion of which, he is carried on from one degree of advancement to another' (Millar 1990: 3). This unique propensity for progress is grounded in human nature.[18] In Smith's terms:

> The principle which prompts to save, is the desire of bettering our condition, a desire which, though generally calm and dispassionate, comes

with us from the womb, and never leaves us till we go into the grave. In
the whole interval which separates those two moments, there is scarce
perhaps a single instant in which any man is so perfectly and completely
satisfied with his situation, as to be without any wish of alteration or
improvement, of any kind.

(WN: 341)

This belief in the potential for social progress is a result of the Scots' belief
that individuals learn by experience. Humans are always progressing on an
individual level in the sense that they are always acquiring experiences
through which they can form expectations. But humans are social beings
who are highly subject to habit formation, socialization and conventional
behaviour. For this reason, the Scots believed that humanity is progressive
on both an individual and a species level (ECS: 7). On a species level this
progress is dependent on the development of conventions in reaction to circumstances that are transmitted to the next generation. In brief, if history is
to be viewed as the progress of the species, then that progress is in the extension of human experience and the development and retention of human conventions and institutions created to 'deal' with that experience. Progress is
the growth of the cumulative sum of human experience, and as all knowledge for the Scots was based on experience, progress is equally the growth
of cumulative human knowledge.[19]

Progress is at base the growth and retention of human knowledge drawn
from experience.[20] As Ferguson puts it: 'the history of every age, when past,
is an accession of knowledge to those who succeed' (ECS: 33). This progression of human knowledge is the basis for the progression of human institutions, and as a result 'industry, knowledge and humanity are linked together
by an indissoluble chain' (EMPL: 271). In practical terms this means that
each human society has a progress of its own, a national progress, which is
but a part of the progress of the species as a whole.[21] The gradual development of human institutions in reaction to circumstances and based on past
experience and example leads to a progress in all areas of human endeavour.
Thus government and law have a progress that is observable by such features
as the development and refinement of notions of property. Progress on a
national and social level is in reality the result of advance in a vast area of
interconnected human institutions. As human knowledge broadens and
deepens every area of human activity is subject to refinement. As a result a
'polished' or civilized nation is marked by refinement in a number of fields.
The growth of knowledge and of the arts and sciences refines human behaviour, civilizes and 'softens' (EMPL: 170) their tempers and interactions. It
changes attitudes towards society and other individuals. For example there is
a progress of opinion that alters the moral and political outlook of humans
as society progresses.

The Scots' analysis of progress does, however, lay stress on its universal
aspects, on those which allow a science of progress as part of the 'science of

man'. As we have seen the Scots' stadial theory is based on the universality of the need for subsistence. As a result Smith is able to trace all progress, in its origins, to this concern. The desire to secure subsistence, to cater for the 'three great wants of mankind ... food, cloaths, and lodging' (LJP: 340) is the root of almost all human art and science (LJP: 337). Such a universal concern forms a great part of the concern of each member of a social group: the desire for survival and sustenance being a core aspect of every human's interest. It is because of this, Smith argues, that the progress of material goods may be traced to a universal self-interest displayed by each human. Material progress may be a result of the growth of knowledge, but its origin is traceable to a mixture of self-interest and a desire for subsistence (LJP: 489). As we saw when we considered the Scots' notions of the immediacy of savages (in connection with the growth of understanding in science and the concept of property) the concern for subsistence consumes human attention when it is hard to come by. But when subsistence is safely secured humanity's attention is turned to other areas and industries. It appears that material progress is a necessary 'requisite' (LRBL: 137) for intellectual and artistic progress: that some measure of security and ease is required before humans are able to develop their understanding of the arts. As security and law develop from barbarity, through habitually accepted conventions grounded in utility, and government becomes accepted, so learning advances (EMPL: 115–16).

In terms of the central concern of subsistence we have seen that humans use their knowledge on the physical environment to provide for subsistence. In this sense individuals seek 'useful knowledge' (ECS: 171): we may learn a great deal from experience but our attention will, in a great measure, be drawn to that knowledge of which we are able to make use. Knowledge is not simply pursued for its own sake, rather its significance arises from the use to which it is put by those who hold it. This is the basis of the Scots' rejection of scholasticism and learning in retreat. Knowledge is acquired through experience and not from abstract reasoning. For this reason, as we saw, the Scots' concept of science is guided by the utilization and analysis of experience rather than by abstracted rationalism. It is because science and knowledge are based on experience that the Scots believe the progress of knowledge is based on the examination and use of past experience. By learning from our own prior experience and by observing that of others we are able to judge better in our future actions. The growth of knowledge, then, is the growth of experience, of human reactions to circumstances.

Moreover, such knowledge can exist in forms that are not immediately explicit. Habit and custom for the Scots were forms of experience-based knowledge: knowledge which is non-verbalized yet vital to the success of our actions. As Smith notes: 'And from all those volumes we shall in vain attempt to collect that knowledge of its [agriculture] various and complicated operations, which is commonly possessed even by the common farmer' (WN: 143). The complexity of the knowledge held in the form of habits on a social level is such that it is both hard to assess and difficult to encompass fully. Its basis is indeed experience but it is individual experience. So when the Scots argued that the growth of experience is necessary for progress they are aware that such experience is experienced by individuals: that though progress is the growth of cumulative experience, the medium of that progress is the experience of specific individuals. It is for this reason that Hume notes that the growth of the cumulative sum of human knowledge does not lead us all to become geniuses (EMPL: 210). Rather knowledge is diffused in line with the individual experience of circumstances. We each have a unique individual sum of knowledge, but that sum is limited by our individual experiences; and this kind of tacit or habitual knowledge reinforces the Scots' point about the customary nature of human knowledge and understanding. Habit and not reason is our guide in the conduct of our everyday life in the sense that our habits embody ways of acting and thinking which prove useful to us in the conduct of life.

We have already seen that Smith viewed population size as an indication of progress and also that he considered it to be the driving force behind the advance in modes of subsistence. However, we have also seen that population pressure is not the means for that advance; rather the means lies in the acquisition of new knowledge to support that population. Progress generates population. It also means that, subsistence having been secured, there are a larger number of people who are able to apply their attention to the development of the various arts of human life. In other words, the greater the population, the greater the scope for the advance of cumulative knowledge and, as a result, the greater the advance in material production and in other human institutions such as government and law. It is the gradual extension of knowledge of the means of subsistence that allows increases in population. Indeed the arts that improve the provision of subsistence become the origins of other areas of human art. Trade also brings societies into contact and allows a flow of ideas, an acquisition of knowledge gleaned from the experience of others, which spurs progress in society (EMPL: 328). The retention and transferral of knowledge is vital to the sustained material position of a society.

Progress is not just the increase in cumulative knowledge, but it is also, in Dunbar (1995: 317) and Dugald Stewart's (1793: 311) terms, 'the diffusion of knowledge' within a society and between societies. The centrality of the thirst for knowledge has already appeared in our examination of Smith's theory of the motivations behind the practise of science. There we saw that humans have in their nature a desire, indeed an emotional need, for systematic knowledge to calm their minds and allow them to proceed about their lives. We also saw that the acquisition of knowledge increases our curiosity. So it is that progress leads to new fields of enquiry as our attention is piqued by new phenomena and 'gaps' in established systematic knowledge. Knowledge itself is a prompt to action and further enquiry leading to a situation of ever-growing complexity and extension of the cumulative sum of human experience. The progress of knowledge on a social or cumulative level is based on the development of experience, thus each stage is based on that before it. Cumulative knowledge is a 'chain' of development that draws upon and refines historical precedent. For this reason it is often difficult to trace the origins of a particular art or practice because it represents the cumulative result of countless modifications and innovations (ECS: 163). Social progress, the cumulative sum of human knowledge, requires that knowledge, once gleaned from experience, is preserved and transferred rather than being lost at the death of the individual who held it. This is why, as we saw, the human species has a progress (through history) greater than that achieved by any specific individual.

The progress of knowledge, and progress more generally, does not occur by sudden leaps of understanding made by specific individuals. Just as the Scots demoted purposive rationality in their attacks on great legislators, so too are they aware that this is not how societies experience progress. The Scots, and Ferguson in particular, deploy organic and biological metaphors to underline the gradualist nature of social change.[22] Indeed Hume argues that our proclivity for habit formation to stabilize expectations and ease the mind, implies that we are emotionally prejudiced against sudden change (THN: 453) – just as it was previously argued that a legislator would find it hard to persuade a people to follow his innovations if they departed radically from previous practice. The whole of the Scots' 'four stages' schema is posited on the notion of such evolutionary, spontaneous order, approaches to social change. This distaste for sudden change does not, however, lead the Scots to believe that humans are hopelessly conservative, for if this were the case then no concept of progress would be possible. Rather we are attracted by 'novelty' (EMPL: 221) so long as it does not occur in a sudden, wonder-inducing manner. Human inventions, the Scots believed, were subject to constant change (THN: 620), but this change occurs slowly and over an extended period of time. Change and progress in human society occur in an evolutionary and not a revolutionary manner.

Another aspect of the gradual nature of social progress is that such change is often 'insensible' (WN: 343–4; THN: 256). That is to say that it occurs so slowly that we do not notice it until it has happened. Ferguson makes the point well:

> But he does not propose to make rapid and hasty transitions; his steps are progressive and slow; and his force, like the power of a spring, silently presses on every resistance; and effect is sometimes produced before a cause is perceived; and with all his talents for projects, his work is often accomplished before the plan is devised.
>
> (ECS: 12–13)

Social change and the progress of knowledge are not only often unnoticed as they gradually occur, but are also often unintended.[23] Our reactions to specific circumstances and the knowledge we draw from repeated experience of them occur in a moment, we react to the specific conditions and then pass on absorbing the knowledge and adapting our practice in the light of it. Mankind progresses by 'gradual advances' (Millar 1990: 4): rational deliberation of long-term advantage plays little role here. The Scots' analysis of the origins of government and property shows that it is limited attention to the moment, repeated often, which leads to the development of social institutions.

The gradual growth of knowledge is the result of a chain of inventions drawn from every aspect of human experience. Humans are always experiencing things and they are always in search of experience, they are constantly adapting to new circumstances as they arise. As a result:

> Those establishments arose from successive improvements that were made, without any sense of their general effect; and they bring human affairs to a state of complication, which the greatest reach of capacity with which human nature was ever adorned could not have projected; nor even when the whole is carried into execution, can it be comprehended in its full extent.
>
> (ECS: 174)

The means by which knowledge is retained and transferred is thus of equal importance for society as its initial discovery or development. Without retention and transferral society would not have a progress distinct from that of the individuals that compose it.

The role of government

Having discussed the Scots' analysis of the evolution of political institutions we now pass on to examine their analysis of the role, or proper function, of government in a commercial society. In broad terms the 'great object of policy' (ECS: 139) for the Scots is the securing of subsistence and the advance of the people. Smith, in discussing the role of government in his system of natural liberty assigns three duties to the sovereign:

> first, the duty of protecting the society from the violence and invasion of other independent societies; secondly, the duty of protecting, as far as possible, every member of the society from the injustice or oppression of every other member of it, or the duty of establishing an exact administration of justice, and, thirdly, the duty of erecting and maintaining certain publick works and certain publick institutions, which it can never be for the interest of any individual, or small number of individuals, to erect and maintain.
>
> (WN: 687–8)

Protection from external threat, maintenance of order through the justice system and the provision of certain public works encompass the whole of the scope of government action. Drawing on our previous discussion of the evolution of government we see that though protection from external threat played an instrumental role in the Scots' analysis of the rise of chiefs, the major benefit of the emergence of government is the enhancement of the delineation of property and the administration of justice. Government, as Hume argued, is an invention to execute justice: the system of justice being necessary for society and civilization itself to flourish. Government evolved to settle disputes over property and to protect the individuals holding that property. As we noted above the chief utility of property delineation is that it brings stability to society, and so we may conclude that the internal role of government is to provide stability, and security to individuals. This internal stability is also a feature of the external defence offered by government in the sense that it creates the 'space' in which the internal order can be kept stable. The provision of impartial justice becomes a key step in creating a stable society and, more particularly, in allowing the development of trade and commerce.[24] Such peace is essential to the gradual development of trade, which is the basis of specialization, and to the growth of the cumulative sum of human knowledge that we have identified as progress. Civilization depends on peace, with the development of knowledge, commerce and the progress of manners all dependent on the internal stability of the society.

This having been said, though government is vital in a society which is progressive, the actions of particular governments can often be severely detrimental to the progress of a nation. Smith argues that progress sometimes occurs in spite of the interference of governments. The 'injustice of human laws' (WN: 378) can retard the economic progress of a nation by the pursuit of policies that pervert the operation of the system of 'natural liberty' which facilitates economic progress. It is because of this that the Scots air doubts over the competence of governments to act effectively in areas of commercial and economic concern. Smith argues:

> The stateman, who should attempt to direct private people in what manner they ought to employ their capitals, would not only load himself with a most unnecessary attention, but assume an authority which could safely be trusted, not only to no single person, but to no council or senate whatever, and would nowhere be so dangerous as in the hands of a man who had folly and presumption enough to fancy himself fit to exercise it.
>
> (WN: 456)[25]

The Scots believe that the role of government in economics and the promotion of industry is restricted to the provision of that stability and peace which allows individuals to pursue their own economic concerns and mutually to adjust to each others' actions. Smith in particular argues that the spontaneous order of free trade is a more efficient medium of economic progress than any system burdened by the interference of governments (however well intentioned) (WN: 687).[26] The reasons for the Scots' doubts over the ability of governments to act effectively in economic matters are, once again, related to their epistemological concerns. We have seen, from the Scots' critique of great legislator theories, that they believe that it was impossible that one 'man of system' (TMS: 233) could effect a complete reform of the social system. When these two elements are combined we begin to see why the Scots have concerns about one individual or group of individuals being moved to attempt radical intervention in the organization of society. The epistemological difficulties which move the Scots' concerns in this matter, what has been referred to previously as the 'demotion of purposive rationality' (Berry 1997: 39), lead them to warn against the possible abuse of the institution of government by those who believe that their vision qualifies them to enforce certain policies upon a society.[27]

These doubts over both the effectiveness and desirability of systematic attempts to reform society and mankind through the policy of government lead the Scots to call for certain constraints to be placed on the actions of governments.[28] An example of the Scots' argument about the limited ability of governments to act to secure desired policy goals in an effective manner is to be found in Ferguson's discussion of population. He argues that population growth is an unintended product of the self-interested action of individuals to improve their situation and supply for their subsistence, noting that no government policy to encourage population has ever affected it to such an extent as this principle. He ends by cautioning against future attempts by government to encourage population growth arguing that: 'A people intent on freedom, find for themselves a condition in which they may follow the propensities of nature with a more signal effect, than any which the councils of state could devise' (ECS: 135–6).

The Scots argue that the systems of mutual interaction and adjustment which humanity has evolved through time are more efficient in such matters. This argument ties in with their view that governments ought to proceed according to general rules. Under a system governed by general rules (a system of the rule of law) the decisions requisite to control, say, prices would require a level of arbitrary action and local knowledge beyond the comprehension of the legislator, while at the same time breaching the principle that he ought to act in the form of generalized principles.

Thus far we have dealt chiefly with the role of government in the provision of justice and that stability which allows social interaction and commercial advance. What we now pass to are those other responsibilities of government that Smith groups under the heading of police. There are, beyond the concerns of internal and external security, certain activities which may be deemed public goods, the provision of which it falls to government to ensure. Smith believes that:

The third and last duty of the sovereign or commonwealth is that of erecting and maintaining those publick institutions and those publick works, which, though they may be in the highest degree advantageous to a great society, are, however, of such a nature, that the profit could never repay the expence to any individual or small number of individuals, and which it, therefore, cannot be expected that any individual or small number of individuals should erect or maintain.

(WN: 723)

The relatively low profits attainable from the provision of these services leads to a situation where price signals and incentives are insufficient to promote private provision. They are, however, in many cases (such as roads and transit) requisite for the pursuit of trade as a whole (WN: 815). Smith believes that these police expenditures are the proper field of government activity; that they provide the stage upon which commercial activity is undertaken. Like justice and security they promote the advance of a society in the sense that they are conditions that foster industry and ease trade. Public goods exist, to a certain extent, below the market. That is to say they facilitate its operation but are not determined by the same principles and incentives that guide wider commerce. The expenditure of police is necessary to the smooth operation of society, to the effective operation of the wider market, and as such must be provided by the government.

We have seen that the Scots apply a spontaneous order approach to the explanation of the origins and development of law and government. Grounding their approach on the underlying universal characteristics of sociability and order-seeking, they provide a conjectural history of the gradual evolution of the institutions of law and government as an unintended consequence of the reaction to the circumstances that groups of humans find themselves in. The analysis dwells on the conventional development of the acceptance of authority from a sense of its utility in conflict resolution. Their 'four stages' schema stresses the significance of the development of different modes of subsistence to the form that the institutions of a society are likely to take, while at the same time highlighting the role of the growth of knowledge in the Scots' conception of social progress. The tasks that are considered the proper function of government are similarly related to the desire to provide stability of expectations in order to facilitate the emergence of benign spontaneous orders, particularly those related to the development of trade, and thus represent another facet of the invisible hand within the society.

5 The science of political economy

The division of labour

As we have seen, the use and transferral of knowledge is one of the key elements in the Scots' understanding of the nature of social progress.[1] This is clearly exemplified by Smith's analysis of the concept of the division of labour. As Smith famously begins the *Wealth of Nations*: 'The greatest improvement in the productive powers of labour, and the greater part of the skill, dexterity, and judgement with which it is any where directed, or applied, seem to have been the effects of the division of labour' (WN: 13). It is through his analysis of the division of labour that Smith explains the phenomenon which results in a situation where the most ordinary labourer in a commercial society has more material resources, is better provided for, than the monarch of a savage or undeveloped country (WN: 24). But more than this, though Smith's famous example of the productive improvements of the division of labour in the manufacturing of pins (WN: 14) graphically illustrates the material benefits of the process, he was also keen to stress the social implications of the division and the wealth which it generates. Indeed Smith goes so far as to state that civilization itself is dependent on the division of labour: he writes: 'In an uncivilized nation, and where labour is undivided' (LJP: 489). This juxtaposition of civilization with the division of labour indicates how central the concept is to his theory of society and social progress.

The division of labour is central to civilization but it is the result of a process of unintended consequences. As Smith would have it: the division of labour 'is not originally the effect of any human wisdom, which foresees and intends that general opulence to which it gives occasion' (WN: 23); rather 'it is the necessary, though very slow and gradual consequences' (WN: 23) of the interaction of human nature with the circumstances in which it finds itself: in brief it is the result of the growth of experiential knowledge (LJP: 570–1). Again and again Smith stresses that the division of labour is not the product of purposive or deliberative human action guided by rational analysis. He claims that: 'No human prudence is requisite to make this division' (LJP: 351), and that 'This division of work is not however the effect of any human policy' (LJP: 347).

The division of labour is not the product of deliberative human action on a social level, a commercial society is not foreseen and planned; rather it arises gradually because of certain 'natural' forces (WN: 278). The division leads not only to technological and material advance, but also to increased, and increasing, interdependence. A division of labour is dependent on an inclination and capacity to trade, and it is here that Smith finds the unconscious spring that allows the development of the division and ultimately of civilization itself. Smith notes the significance of the fact that humans are the only animals which trade (LJP: 352), that while other animals co-operate to achieve ends – two greyhounds running down a hare is his example (WN: 25) – this is the result of 'the accidental concurrence of their passions in the same object at that particular time' (WN: 25–6). This, and other animal behaviour – such as fawning puppies – is not the same as conscious trade and exchange. 'Nobody', Smith writes, 'ever saw a dog make a fair and deliberate exchange of one bone for another with another dog' (WN: 26).

Trade is a uniquely human activity. Smith illustrates this by noting that, although unaware of the concept of the division of labour (LJP: 335, 521), savages nonetheless practise the exchange of surplus that is the origin of the phenomenon. This for Smith indicates a 'disposition' (WN: 27) or a 'propensity' (WN: 25) in human nature to 'truck, barter, and exchange one thing for another' (WN: 25). The division of labour is an unintended consequence of this facet of human nature: humans seek to exchange what they have for what they want. The system is not developed intentionally, the propensity has, in Smith's words 'in view no such extensive utility' (WN: 25). The concept of utility involved is far more localized and short term. As Smith put it:

> Twas thus a savage, finding he could by making arrows and disposing of them obtain more venison than by hunting, became an arrow maker. The certainty of disposing of the surplus produce of his labour in this way is what enabled men to separate into different trades of every sort.
>
> (LJP: 351)

This division is the result of self-interest, the initial exchanges being based on a desire to satisfy individual wants, and the eventual decision to specialize resulting from the observation – drawn from experience – that these needs are better provided for as a result of concentration on one productive activity which may then be traded. Individuals become increasingly interdependent as a result of such specialization, and humans come to depend on each other to supply for their needs through the medium of trade.

A further important factor is that trade is based on the interaction of individuals seeking to fulfil short-term utility. Thus experience soon teaches them that the quickest and most efficient means of securing the co-operation and trade of others is to appeal to their self-interest. As Smith famously states: 'It is not from the benevolence of the butcher, the brewer, or the baker, that we expect our dinner, but from their regard to their own interest' (WN: 26–7). The maker of arrows appeals to the self-interest of the hunter. Hunters will no longer be required to produce their own arrows if they can exchange their surplus for those produced by another, and that surplus will grow as a result of the time freed up from arrow making which they can then devote to more hunting. The benefits of specialization are dependent on the inclination to trade.

Though Smith provides little in the way of explanation behind the 'trucking' principle he does make one revealing aside which links it with his conception of sympathy. He argues:

> If we should enquire into the principle in the human mind on which this disposition of trucking is founded, it is clearly the naturall inclination every one has to persuade. Men always endeavour to persuade others to be of their opinion even when the matter is of no consequence to them.
>
> (LJP: 352)

This desire to persuade is clearly related to Smith's argument in the *Theory of Moral Sentiments* about the human emotional need for the approbation and approval of others. In terms of trade this principle is compounded with the desire for subsistence: with utility teaching humans from experience that the surest way to secure the co-operation of others, the surest way to persuade others to assist in the satisfaction of your wants, is to trade – to persuade by bargain and exchange. As a result the wider the scope for trade the wider the scope for specialization. The greater the number of potential trading partners, the greater the incentive to specialize. In scattered communities 'every farmer must be butcher, baker and brewer for his own family' (WN: 31), interdependence is not possible because of geographic isolation. Specialization is not possible unless a market of sufficient size is available, unless there are enough potential trading partners. The division of labour advances in proportion to the scope for trade: specialization and interdependence lead to increased contact between people, and through trade to a concomitant increase in population centralization.

Distinct industries or employments develop with this specialization, with the original suggestion of career path being an apparent 'natural' talent for a particular form of labour. But though this forms the basis of the impetus to specialize in a particular task in the early stages of the division, we see that, as specialization advances, the notion of 'natural' talent begins to take a back seat. What instead comes to matter is the specialized knowledge that the individual acquires from devoting their attention to a particular profession. Smith seeks to make it clear that he is not arguing that differing natural attributes and inherited faculties are the basis of specialization and the benefits which arise from it. Rather that skills and attributes are acquired as a result of the division itself. He says:

The difference of natural talents in different men is, in reality, much less than we are aware of; and the very different genius which appears to distinguish men of different professions, when grown up to maturity, is not upon many occasions so much the cause, as the effect of the division of labour.

(WN: 28)

He follows this by asserting that: 'The difference between the most dissimilar characters, between a philosopher and a common street porter, for example, seems to arise not so much from nature, as from habit, custom and education' (WN: 28–9). The point which Smith is trying to make is not so much that natural abilities are unimportant, but rather that under a system of specialization the differences brought about by application to a particular field of work are a more decisive factor in explaining the broad variety of different individuals and their respective skills and sums of knowledge.

However, just as we require trade to allow specialization, so too does the increase in specialist skill which it produces depend on the interaction of individuals through trade. Smith dwells on the fact that dogs, who display far greater natural differences in behaviour, are unable to exploit these differences in the interests of the species precisely because they do not possess the propensity to trade.

Among men, on the contrary, the most dissimilar geniuses are of use to one another; the different produces of their respective talents, by the general disposition to truck, barter, and exchange, being brought, as it were, into a common stock, where every man may purchase whatever part of the produce of other men's talents he has occasion for.

(WN: 30)

Having discussed the factors which lie behind the separation of arts and professions, and examined how this is related necessarily to the notion of trade, Smith then goes on to examine the division of labour as it develops within the various, now delineated, industries and professions. Smith lays down three reasons why the division of labour produces productive benefits when introduced to the internal operation of a particular productive industry. He attributes this:

first, to the increase of dexterity in every particular workman; secondly, to the saving of time which is commonly lost in passing from one species of work to another; and lastly, to the invention of a great number of machines which facilitate and abridge labour, and enable one man to do the work of many.

(WN: 17)

What is immediately striking about these three explanations, given what we have already seen about the role of specialized knowledge in the separation of arts and professions, is that two of them, the first and the third, refer to benefits which are the result of improved skill and knowledge. The second is categorically different, referring instead to the actual nature of the working environment. The second explanation is also the weakest and least productive of the three. While Smith is right to note this difference between simpler models of production where a craftsman works on each stage of production and the more compartmentalized chain of production under the division of labour, the savings of time attained by the prevention of 'sauntering' (WN: 19) surely cannot be considered to be of so great an improving force as the increase of dexterity and the invention and use of machines. Indeed, once the division of labour has first been introduced to an industry it is doubtful as to how great a difference the elimination of time-wasting in the change between functions will truly be. On the other hand the role of the other two explanations of the productive powers of the division of labour are not so limited and may fairly be said to be of constant relevance as each industry progresses. It is the increased dexterity of workers and the invention of machines which are the truly progressive elements of the division of labour.[2] Further, the first and third explanations are logically linked together in Smith's argument. The initial explanation for the increase in dexterity is the simplification of the task in hand: 'the division of labour, by reducing every man's business to some one simple operation, and by making this operation the sole employment of his life, necessarily increases very much the dexterity of the workman' (WN: 18). By confining an individual's attention to a simple field the division of labour focuses attention and creates a specialist whose skill and knowledge of this operation allow them to perfect it to levels beyond the power of a generalist. As we saw earlier in our discussion of science and 'gap-plugging', specialists are more intimately familiar with and focused upon their particular field; as a result they are able more easily to apprehend 'gaps' in that system or operation, and are able to apply their focused knowledge to plug those 'gaps'.[3] What emerges from this is the notion of an occupation as a 'study', perhaps the first definition of the idea of human capital.[4] As Smith puts it:

> Those talents, as they make a part of his fortune, so do they likewise of that of the society to which he belongs. The improved dexterity of a workman may be considered in the same light as a machine or instrument of trade which facilitates and abridges labour.
>
> (WN: 282)

Such knowledge gained from experience and repeated exposure to a particular field is held within the minds of the individuals concerned. For this reason such knowledge often takes on a non-verbalized, tacit or habitual form. Ferguson notes this when he argues:

Accessions of power in us are sometimes termed skill, and consist in the knowledge of means that may be employed for the attainment of our end: they are also termed a sleight or facility of performance; and are acquired by mere practice, without any increase of knowledge. The first is the result of science; the second is the result of habit. And there are few arts or performances of moment, in which it is not requisite that both should be united.

(Ferguson 1973 vol. 1: 227–8)

It is not simply the possession of specialized knowledge that counts, but also the manner in which it is exercised: the skill we have in utilizing our knowledge. For specialization to work what is required is that specialists are proficient in their own field; that they are able to act in a relatively efficient manner on the objects that are the focus of their attention. Indeed Smith notes that one of the advantages of such specialization is the scope which it allows for the conduct of experimentation by informed practitioners: a process which is vital to the progress both of knowledge and of wealth. As specialization advances more people become specialists in the same field resulting in a situation where, according to Smith: 'More heads are occupied in inventing the most proper machinery for executing the work of each, and it is, therefore, more likely to be invented' (WN: 104).[5]

The productive benefits of specialization are related to experience and to the acquisition of specialized knowledge. Smith links this specialization to his third explanation. He argues: 'Men are much more likely to discover easier and readier methods of attaining any object, when the whole attention of their minds is directed towards that single object' (WN: 20).[6] This argument is further underlined when Smith admits that some great mechanical innovations are not the result of the experiences of workmen, but rather are the product of those who specialize in making machines: a further example of the benefits of specialization (WN: 21). The restriction of attention to one field of study, or occupation, naturally increases the scope of the observations that may be made in that field by any one individual.

The process of specialization, however, does have limits and equally gives rise to potentially serious problems. Individuals are restricted by the limits of their mental functions, they are only capable of processing so much knowledge and even then the nature of human knowledge, founded as it is on habitual association and experience, is imperfect. As a result the concentration of our attention on one field of study, though efficient, naturally restricts our ability to process knowledge from other fields. For this reason it is inevitably the case that we cannot fully comprehend the details of the fields of other specialists as these lie outside our experience.[7] Hume is also quick to note that specialization does not render us all experts, or geniuses, in our chosen field. Rather we are merely specialists proficient in the skills required for the operation of our own field but necessarily constrained in our understanding of the wider system within which our field operates (EMPL:

210).[8] One danger of this process is that specialists may acquire tunnel vision, focusing their attention on one field and blinding themselves to the significance of other fields of study. This results in a situation where the 'ordinary person' only acquires experience of other fields 'second hand', through the teaching of others or observation (LJP: 574). There is, then, a potential danger, highlighted by Ferguson (ECS: 32, 206–7), that concentration on a specialist area of study leaves us ill-equipped for involvement in other specialists' areas: or that our proficiency in one field is bought at the expense of our ability to interact in vital social activities.

Knowledge specialists, as we saw in our examination of the division of labour, must interact for their specialized knowledge to be useful. Moreover, specialists become dependent on the knowledge and labour of others to an extent that interaction and trade become vital. We become dependent on the skill and knowledge of others and, as individual fields of experience are focused further and further to reap the benefits of close study, so society becomes increasingly complex, experience increasingly diverse, and interdependence gradually greater and greater. Knowledge is indeed increased in its cumulative sum, but it is also diffused among an ever-wider field of specialists. This cumulative growth in knowledge, which we have observed before, shows us that knowledge itself is a chain of development conducted through the medium of specialists. Specialists build on the work of those who have gone before them. Moreover, this 'accumulating advantage' (ECS: 199) from specialization depends upon the focus and skill of each specialist in his own field. The gradual efforts of individual specialists to exert themselves in their own field, and in their own interest as we shall see below, benefit the whole of society by increasing the stock of cumulative knowledge. What becomes clear is that specialization reinforces the notion that the knowledge of the whole of a society exceeds that of its discrete members. But specialization also encourages the growth of the sum by focusing attention on individualized fields leading to a development of proficiency in them which benefits all through trade and interdependency. All of this is posited on the interaction of the individuals who compose a society: interaction and trade are vital if specialized knowledge is to be gathered or utilized to the benefit of all. If cumulative social knowledge is to mean anything, then there must be social interaction through which to make use of it. In a complex commercial society knowledge must be transferred, indeed, as Smith put it, knowledge must be brought into a 'common stock' (LJP: 573) by trade.

Interaction to utilize individualized specialist knowledge is essential to the progress of the cumulative sum of human knowledge that we have identified as the basis of the Scots' conception of progress. Just as Smith notes the vital role of the desire to trade, arising from the propensity in human nature to truck and barter, in allowing the development of the division of labour so, it becomes clear, is trade also vital to the development of the specialist knowledge which underlies the process. Hume argued that the stability of property and the recognition of its free transfer by consent laid the foundations for society while at the same time creating the conditions for the development of trade. Moreover, just as stability of property promotes peace in a given society so trade, by bringing people into peaceful contact, promotes civilization and encourages the exchange of ideas. Commerce, in supplying for the needs of subsistence, becomes the great 'study' of mankind.

But market relationships differ from other forms of human interaction. As Smith noted, to appeal to the self-love of the butcher, brewer and baker is a more efficient means of securing that which we desire, and, as Hume also notes, market relationships differ from friendship and appeals for the sympathy of others (ENQ: 209). The truth of these statements becomes even more salient as trade develops and specialization increases. We become dependent on the skills of others to supply our wants, while at the same time they become equally dependent upon us. Such interdependence grows up to a great complexity as the division of labour advances. As Smith famously asked us to observe:

> the accommodation of the most common artificer or day-labourer in a civilized and thriving country, and you will perceive that the number of people of whose industry a part, though but a small part, has been employed in procuring him this accommodation, exceeds all computation. The woollen coat, for example, which covers the day-labourer, as coarse and rough as it may appear, is the produce of the joint labour of a great multitude of workmen.
>
> (WN: 22)[9]

The example of the labourer's coat indicates the vast web of interdependency which develops as a result of the division of labour; but it also shows how this complexity supplies for our needs in an efficient manner, and in a manner that depends on a market exchange which allows us to depend on people unknown and unrelated to us. The various stages of production interact to serve their own ends, sheep farmers have no inclination or idea that their wool will eventually supply the clothing needs of a labourer. All that concerns them is exchanging their product for their own advantage in order to procure the satisfaction of their own needs. This complex of interdependency, of reliance on the skills and labour of other specialists, is what allows the modern labourer to be better provided for than the African chief (WN: 24). The interdependency of specialists through trade depends, for Smith, on the extent of the market, on the number of possible trading partners. As trade extends it also develops, as Hume notes, geographically distant areas come to enter trading relationships extending the market and the scope for trade still further (EMPL: 299). This trade brings people into contact with new civilizations and their ideas and products, allowing the exchange of ideas as well as goods promoting the further enhancement of knowledge. If the key to the success of the division of labour is the extent of the market

(WN: 34): with the implication being that as the division of labour improves products and the division of knowledge extends the cumulative sum of human knowledge, so the market of the greatest possible extent is a desirable situation for mankind. It is from these principles that Smith launches his argument for the efficiency of free trade and open competition (WN: 362).[10]

An interesting feature of Smith's argument here is his belief that interaction through trade is vital not only to material progress and the progress of knowledge, but also to the progress of manners (LJP: 223, 538). Market exchanges and trade lead to improvements in civility and morality, fostering such 'virtues' as probity and honesty as the 'reputation' of a trader becomes as much a part of their product's attraction as the goods themselves (LJP: 539).[11] Similarly bargaining and contract-making become skills in themselves and encourage specialists to develop in that field (merchants) whose livelihood depends on their proficiency in striking deals (LJP: 494). Underlying the efficiency of the trading relationships which foster the division of labour and the division of knowledge is an assumption about human motivation which has often led to debate around the Scots' theory, especially that of Smith and in particular the relationship between his *Theory of Moral Sentiments* and *Wealth of Nations*.[12] The Scots' analysis of trade, though based at bottom on a 'natural' propensity in human nature, is conducted in terms of self-interested utility maximization.

Self-interest and trade

As we have seen, market relationships are not based on love or sympathy, but rather upon self-love. Humans play on each other's self-regard to satisfy their needs through exchange. The reason for this becomes clear when we consider the Scots' notion of 'confin'd generosity': because sympathetic relationships are based on familiarity, or imaginative familiarity, while market relationships are based on interdependency among a large number of people, trade simply becomes an easy way of interacting without taking the trouble to build the bonding relationships or undergo the imaginative sympathy necessary for appeals to sentiment to be fully effective. As barter is more efficient than emotional appeal in a society where individuals who interact are necessarily ignorant of most of those with whom they interact, self-interest becomes a conventional standard which is easily recognizable and understandable to individuals previously unknown to each other. Harking back to the labourer's coat example, we see that there is quite simply no way in which a web of interdependency so complex as that required to produce the coat could have arisen if the process depended on a direct sympathetic relationship between individuals.

Such appeals to self-interest are related to the Scots' contentions over the need for and effectiveness of general rules. The laws that govern commercial exchange serve the function of allowing a stability of expectation in trading relationships with the honouring and enforcement of contracts becoming a key feature of a trade-based society. General rules of interaction governing trade exchanges do not require any notion of friendship (ENQ: 209): such general rules of behaviour stabilize expectations precisely because they reach beyond sympathy.[13] They allow us to understand each other and to interact without detailed knowledge of each other's situations, and without the need extensively to exercise sympathy. We govern our behaviour by socially generated habitual expectations such as propriety, and such general rules reduce the prospect of uncertainty and ease interaction.

The Scots did not believe that all human motivation could be reduced to a principle of self-love or self-interest, and they took especial care to distance themselves from Mandeville's analysis of 'vice' and his attitude to self-interest.[14] But they are equally forceful in noting that human virtue and morality cannot likewise be reduced to a simple motivation of the opposite impulse of benevolence. What the Scots instead argued was that human motivations are more complex and that they are difficult to reduce to single principles. The Scots refer to humans as having motivations which are both selfish and social, arguing that though self-interest is a strong influence on human behaviour it is not sufficient alone to explain social interrelation and behaviour.[15] Indeed, though they condemn excessive self-regard, they are also quick to note that self-interest is an integral part of human motivations. Smith argues that selfish passions are neither social nor anti-social: they are simply facts about the nature of man just as it is a fact that human nature is sociable (TMS: 40, 172).[16] What is required to understand human motivation and to discern the nature of social behaviour is an acceptance of the reality that extremes of self-interest and benevolence are equally undesirable and unrealistic as explanations for morality. Instead the Scots accept the 'reality' of the situation and seek 'virtue' in the avoidance of extremes of either motivation or passion (EMPL: 161).[17] Motives are themselves difficult to discern as we cannot fully experience the passions and thought processes of another human.[18] This is the source of the Scots' focus on and concern with sympathy. Sympathy is an attempt to understand the motives of others in order that we might be able to understand their actions and make judgements about them. The complexity of human motivations imply that even while engaged on self-interested economic exchange, people are also embedded in a series of inter-relations which do not depend on self-interest for their motivations (ECS: 40–1).

The impulse for self-preservation is observable in all animals and is linked to a natural fear of death (TMS: 13). Humans, like other animals, are greatly concerned with self-preservation and by the nature of their animal frame are required to provide for their own subsistence. We all possess the urge for self-preservation: this is simply a fact of nature that is neither a virtue nor a vice. As we have discussed at length before, the role of commerce and much human industry is to provide for subsistence. As a result of this relation of economic activity to the securing of subsistence, and of a similar link between subsistence and the natural urge for self-preservation, economic activity comes to hold as its chief motivation a principle of self-interest or, more accurately, a desire for 'improvement' (LJP: 487).

It is the desire for self-preservation and a regard to our own interest that prompt us to industry (LJP: 340). Moreover, this prompt leads us to exert our talents to the full and the desire for self-improvement becomes the incentive that drives commercial activity, a feature which sits comfortably with the self-regarding exchange that forms the medium of trade. Idleness, Smith believed, is caused by a lack of incentive (WN: 335), and chief among the incentives that prompt humans to action is their admiration for the conveniences of the wealthy. Once subsistence is secure industry develops around the gratification of vanity and the desire to emulate the rich (WN: 190–3).

Self-interest and self-improvement are powerful motives that encourage human industry and endeavour in a variety of social arenas. There is a very real sense in which individuals' self-interest may be incited, through the institutions that constitute the invisible hand, to encourage them to labour and to utilize their skills in order to benefit not only themselves, but also the public as a whole. In terms of knowledge, if an individual is given incentives to study and increase their understanding this adds to the cumulative sum of human knowledge which moves progress regardless of the motives for undertaking the work.

Perhaps the clearest indication of how the Scots' thought self-interest and the market linked with their unintended consequence approach is to be found in their analysis of historical change: particularly in their analyses of the decline of slavery and feudalism.[19] The Scots were universally opposed to the institution of slavery, but they sought to explain the 'happy concurrence of events' (Millar 1990: 261) that led to the decline of slavery in terms of an unintended consequence argument. They stress that claims to a decisive, intentional role in the abolition of slavery, such as that made by the Catholic Church, are inaccurate and unrealistic.[20] Instead, they argued, the fall of slavery and the rise of emancipation were brought about as an unintended consequence of human interaction. The institution of slavery died out gradually as the result of a 'natural progress in manners', it fell into disuse before any legal move was made to outlaw it (Millar 1990: 263, 278). The Scots' reasoning behind this argument is related to the gradualism and evolutionism of their historical analysis. They believed that slavery passed out of favour because it failed to provide sufficient incentives to encourage industry on the part of slaves (Millar 1990: 250–1, 264, 267, 282). Slavery hinders industry because a slave has no share in the product, or profit, of their labours, and as a result has no incentive to maximize their efforts.[21] Slavery gradually falls out of use as slave-owners realize the gain in productivity to be had by liberating slaves and treating them as dependent tenants or employees with some share in the product of their labour. Slavery is uneconomic, and as awareness of this grows it falls out of favour as a system of economic organization (LJP: 454, 580). Slave-owners become landlords and employers as they realize the benefits to be gained by exploiting the self-interest and desire for self-improvement of their dependants. Such appeals to self-interest as an incentive are more economically efficient than the abject dependence of slavery (WN: 387–8). The slave system's inefficiency is highlighted not only by the disincentive of labour which it produces amongst slaves, but also by the disincentive it provides for slaves to improve their skills and knowledge by application to various trades. As Smith notes, slaves are 'very seldom inventive' and 'all the most important improvements' in machinery or the division of labour have been the discoveries of freemen (WN: 684). As slavery declines and freedom advances so industry and commerce begin to develop in an efficient manner, harnessing the self-interest of freed-men to promote industry. Freedom grows along with commerce: the system of tenancy replaces that of slavery and there is a gradual increase in equality in society (LJP: 195, 391). Slaves, who were by their nature unequal with freemen, acquire the same legal status as them. This process of emancipation and the exploitation of self-interest leads in turn to a growth in productive output and to a growth in trade.

The Scots apply a similar system of analysis, focusing on unintended consequences and self-interest, to their study of feudalism. They undertake a historical analysis of the feudal era which is grounded, especially in the case of Smith and Millar, within their stadial theory of historical development. The Scots analyse both the rise and fall of the feudal system in terms of unintended consequences. The rise of feudalism, Smith argues, destroyed the nascent system of trade that had developed after the fall of classical slavery (LJP: 248), but its advent was the result of the interaction of various features related to the system of economic production. The feudal system was, in the Scots' view, the product of a balance of powers between the various eminent nobles of the nation.[22] This balance, or spontaneous order, evolved gradually as landlords became aware of the balance of powers within each particular locality (WN: 402). Moreover, as Millar (1990: 188, 197) argues, this was not a conscious balance. The unintended consequences that produced the stable feudal system produced a system of law that recognized the balance of power between feudal lords and between the lords and the sovereign. Each noble acted from self-interest to preserve his position, but the nobles gradually developed a group interest to defend their feudal rights against the power of the crown, and this balance of nobles and monarch characterizes the feudal system (EMPL: 17).[23] Feudalism gradually developed a customary and legal framework, grounded on this balance of powers and the economic system of dependent tenants. The feudal system was based on the concentration of property, with the feudal lordship over a geographic area being the basis of a lord's power. As a result it became vital to preserve the integrity of property in land. For this reason the feudal system was particularly characterized by its emphasis on the legal concept of primogeniture as a means of securing the maintenance of power by preserving an estate intact at the time of inheritance (EMPL: 413; LJP: 56–7). The need to preserve the integrity of feudal estates also led to the development of such legal features as entailed legacies (WN: 384) where ownership of the feudal right and enjoyment of the power which went with it depended on the preservation of the estate and its value (LJP: 70).[24]

The Scots believed that feudalism arose as the result of a process of unintended consequences, but they also held that this was how the system passed out of existence. Under feudalism the landlord's sole aim was to protect his position of influence by protecting the extent of his estate. His prestige was measured not only by the extent of his land but by the number of dependants who worked this land for their subsistence and who were thus dependent on the feudal lord for their survival. Under the feudal system the powerful have no other means to exercise their wealth than in maintaining dependants: their spending is limited simply by the fact that there is nothing to buy (LJP: 50). The feudal system began to decline only when objects arose which the feudal lords could buy. In other words, when some individuals began to specialize in the production of non-essential, 'luxury' goods, the landlords suddenly acquired an object upon which to expend their incomes (LJP: 262–4). By seeking to satisfy their desire for these non-essential goods as status symbols the feudal lords began to lose their focus on the importance of the concentration of property. Trading land for money in order to fund their taste for the luxurious, for goods that they themselves could consume and enjoy, the lords unwittingly destroyed the very basis of their power. The feudal system was destroyed as an unintended consequence of the self-interest of the feudal lords and the consequent rise in specialized labour to meet this demand. Smith describes this in the following terms:

> But what all the violence of the feudal institutions could never have effected, the silent and insensible operation of foreign commerce and manufactures gradually brought about ... As soon, therefore, as they could find a method of consuming the whole value of their rents themselves, they had no disposition to share them with any other persons. For a pair of diamond buckles perhaps, or for something as frivolous and useless, they exchanged the maintenance, or what is the same thing, the price of the maintenance of a thousand men for a year, and with it the whole weight and authority which it could give them.
>
> (WN: 418–19)

And so, in Smith's terms:

> A revolution of the greatest importance to the publick happiness, was in this manner brought about by two different orders of people, who had not the least intention to serve the publick ... Neither of them had either knowledge or foresight of that great revolution which the folly of the one, and the industry of the other, was gradually bringing about.
>
> (WN: 422)

The principle of self-interest promotes both the decline of feudalism and the advance of commerce and the division of labour. Feudalism operated on a restricted level of incentive, and once the opportunity to enjoy greater profit arose – both for the lord in the enjoyment of non-essential goods, and for the labourers in the opportunity to enjoy the product of their own labour – the feudal system entered a 'natural' and inevitable decline. Feudal wealth gradually declined as property was sold off to fund the purchase of luxury goods: feudal dependants were freed from their association with a particular lord and left able to practise their increasingly specialized trades for their own profit. The decline of feudalism was characterized by the diffusion of previously concentrated wealth and power through the medium of trade. This diffusion of wealth and power led to a gradual improvement in the position of those who had previously been dependent on a particular lord alone for their subsistence.[25] As the division of labour advanced feudal dependency declined while interdependency through trade increased. The wealthy buy the product of the labour of independent manufacturers and indirectly provide their maintenance (WN: 420). Another unintended consequence of the decline of feudalism was the growth of cities and towns.[26] Newly emancipated serfs began to congregate in urban areas in order to practise specialized labour and enjoy the benefits of trade allowed by a more extensive market.

We have already seen that Smith believed that the division of labour and the division of knowledge arose as an unintended consequence of human action and that the commercial system had evolved gradually through his 'four stages'. This unintended process produces a system of specialization and trade that characterizes commercial societies. We are left with a question: Smith must explain how, given his focus on specialized, localized, self-regarding action, 'the private interests and passions of individuals naturally dispose them to turn their stock towards the employments which in ordinary cases are the most advantageous to the society'? (WN: 630). The phrase that Smith uses to describe such incidents, where self-interested actions and circumstances combine to form benefits for the public good, is 'accidents' (WN: 78, 235). It is in the study of these 'accidents' that Smith seeks the nature and causes of the wealth of nations. The interaction of self-interested individuals possessed of an urge to trade is the arena within which such accidents occur. Smith's point is that these individuals do not have the good of society as their aim, but rather in the pursuit of their own interests produce unintended consequences that are in the interests of society as a whole (WN: 454).[27] Smith is advancing an epistemological argument: our superior knowledge of our own particular circumstances and our desire to exploit them efficiently leads, if successfully co-ordinated, to the most efficient exploitation of the circumstances of the whole of the society. Smith attributes the benign spontaneous order that is produced to the operation of an invisible hand. The invisible hand here is the efficient exploitation of local knowledge in a social context.[28]

The invisible hand

In the Introduction we noted that MacFie distinguishes between the first use of the term invisible hand, in the *History of Astronomy* where it appears as the 'invisible hand of Jupiter' (EPS: 490), and its later appearances in *The Theory of Moral Sentiments* and *The Wealth of Nations*. He believes that in all three cases the hand referred to is that of a Deity, but that the nature of the Deity in question changes from a polytheistic context to a Christian context (MacFie 1971: 595–6). In Chapter 2 we examined the Scots' rejection of explanatory models that relied upon a specific role undertaken by a Deity, building a case for a reading of the Scots that sees them as undertaking a self-consciously secular explanatory inquiry. On this reading of the Scots it makes little sense for Smith suddenly to have recourse to divine intervention in order to explain the results of social interaction. Nonetheless, a number of critics have argued that the invisible hand is indeed a metaphorical description of either the direct intervention of God, or the unfolding of the plan of providence.[29] Where the Scots do make passing reference to providence it generally has little to do with the central explanatory thrust of their arguments and sits oddly with their professed scientific approach. A more plausible reading is that offered by Gray, who notes that Smith's use of the term invisible hand to describe the 'self-adjusting machinery' (Gray 1931: 151) of society is a cover for the fact that he cannot describe in precise detail the nature of the co-ordinative device.[30]

Emma Rothschild has recently advanced the view that the focus on the term invisible hand in much Smith scholarship is misplaced. She believes that Smith viewed it as a 'mildly ironic joke', which he uses in a 'cursory' manner (Rothschild 2001: 116, 118) and, that if taken as a substantive theory, it is distinctly 'un-Smithian' (Rothschild 2001: 123–4). One of her arguments that the invisible hand is 'un-Smithian' is precisely that it has superstitious or religious connotations. She believes that Smith's work is essentially secular in character and that, as a result, the invisible hand sits uneasily with his style of approach. This is congruent with our reading in so far as it rejects the idea that the invisible hand is that of the Deity; however, the conclusion that its possible interpretation as referring to a Deity renders the idea 'un-Smithian' requires further support. With this end in view Rothschild advances three arguments in support of her case.[31] First, she undertakes a literary comparison of Smith's invisible hand to prior uses of the phrase of which Smith may have been aware, in Macbeth and Ovid, and concludes that it carries superstitious or miraculous connotations (Rothschild 1994: 319–20; 2001: 118–21). The Scots, as we have seen above, disliked the notion of miracles and favoured causal explanation; which supports the idea that *if* Smith drew the term from one of these literary sources, then it is not a style of argument with which he had any great sympathy. The second line of argument advanced is that the invisible hand adopts a condescending and contemptuous attitude towards individual intentions (Rothschild 2001: 123). This,

Rothschild argues, goes against the very tenor of the Enlightenment idea of freedom through reason. The idea that we are led by forces other than our reason to produce outcomes that were no part of our intention seems at odds with the notion of Enlightenment (Rothschild 1994: 320–1). This line of argument depends on Rothschild's reading of Smith as being closely related to the French Enlightenment. That he is more properly understood as being situated in a less rationalistic and uniquely British school of thought has been the focus of much of our study thus far. This British school of Enlightenment is partly characterized by its appreciation that macro level outcomes of social interaction often have little to do with the particular rational intentions of the actors involved.[32] Indeed the entire discussion in *The Theory of Moral Sentiments*, that only intentional actions are the proper subject of assessments of merit, reflects a keen awareness on Smith's part that some outcomes are not the product of intention and thus not properly regarded in terms of merit (TMS: 97–103). Rothschild's final argument is that the invisible hand posits a theorist with privileged universal knowledge who is able to identify the hand, while the individuals who are guided by it remain unaware of its operation (Rothschild 2001: 124). This idea is distinctly 'un-Smithian' if it is regarded as referring to some superior exercise of reason on the part of the theorist that would allow him to predict the outcomes of interaction. However, Smith's use of the invisible hand is, as we shall see, as a retrospective explanatory device. The theorist cannot see more than the actors, but with hindsight he can identify the operation of the hand that produces results other than those intended by the actors. The invisible hand is not necessarily 'un-Smithian', and it now remains for us to identify the particular role that it plays in his work.

In the *Theory of Moral Sentiments* the invisible hand appears in a section dealing with the effect of utility on the conception of beauty. Smith argues that the rich in a society are subject to the same physical constraints as the poor, that is to say that their corporeal frames restrict the amount which they can absolutely consume. As a result they are compelled to use their wealth to purchase the product of others' labour, and consequently they diffuse their wealth through society. As Smith would have it:

> They consume little more than the poor, and in spite of their natural selfishness and rapacity, though they mean only their own conveniency, though the sole end which they propose from the labours of all the thousands whom they employ, be the gratification of their own vain and insatiable desires, they divide with the poor the produce of all their improvements. They are led by an invisible hand to make nearly the same distribution of the necessaries of life, which would have been made, had the earth been divided into equal portions among all of its inhabitants, and thus without intending it, without knowing it, advance the interest of the society, and afford the multiplication of the species.
>
> (TMS: 184–5)

Here he is attempting to show that economics is not a zero-sum game. That is to say that the nature of economic interdependence implies that the ownership of more wealth by some does not entail a loss of subsistence by others. The invisible hand here is the mechanism by which a benign spontaneous order, one that is in society's interests in general, can be produced by the self-regarding actions of individuals.

Similarly, in the *Wealth of Nations*, the appearance of the invisible hand is again related to the co-ordination of self-interested action in order to produce benefits for the whole of society. Here the question is the balance of trade. Smith writes:

> By preferring the support of domestick to that of foreign industry, he intends only his own security; and by directing that industry in such a manner as its produce may be of the greatest value, he intends only his own gain, and he is in this, as in many other cases, led by an invisible hand to promote an end which was no part of his intention. Nor is it always the worse for the society that it was no part of it. By pursuing his own interest he frequently promotes that of the society more effectually than when he really intends to promote it. I have never known much good done by those who affected to trade for the publick good.
>
> (WN: 456)

As with Smith's other examples of the role of economic self-interest in historical change the term invisible hand refers to the process, or mechanism, which brings about socially beneficial spontaneous orders from the interaction of self-regarding actors. Whether the result is in the distribution of subsistence, or in the support of domestic industry, the process is the same.

It is for this reason, the efficient exploitation of local and specialized knowledge, that we begin to see why Smith stressed the point that national capital and wealth is nothing more than the sum of the capital and wealth of the individuals who compose the nation (WN: 366). Public and private goods are interdependent, and the desire to improve their position held by individuals is the force that improves the position of the nation as a whole. All of this having been said it becomes clear that, just as the division of labour depends on trade to allow the interdependence of specialists, so trade plays a vital role in the co-ordination necessary to utilize the dispersed knowledge of those specialists. For the cumulative sum of specialist, individuated knowledge to be useful it must be brought into co-ordination. The efficient use of local knowledge by individuals interacts to create social benefits through the medium of the invisible hand. What must now be determined is the nature of that hand.

Markets and prices

We have already discussed how Smith's *Theory of Moral Sentiments* portrays a subjective, inter-personal, generation of moral value; how the desire for approbation and the impartial spectator lead us to 'see ourselves as others see us' (TMS: 110) and consequently to moderate our emotional displays. This, we argued, shows that the generation of moral values is the adaptation to social circumstances of sympathetic beings. Virtues such as prudence and propriety are inter-subjectively generated and become objective in the sense that they become habitual and socially accepted. Value arises from comparison: it is a subjective standard dependent on a comparison undertaken within a specific set of circumstances. In a psychological sense, phenomena such as sympathy and the impartial spectator allow us to understand the actions and motivations of others. We are unable to experience precisely what they feel, but we can, through imagination, place ourselves in their shoes. In terms of economic exchange we know from our knowledge of ourselves that we are self-interested in acquiring the means of subsistence; we then extend this principle to others, of whose motivations we are necessarily ignorant but suppose to be similar to our own, and simple exchange by appeal to self-interest becomes possible. Though the same analysis may be applied to benevolent action this is less efficient as a mode of exchange, as it requires more extensive sympathetic imagination and greater intimacy with those with whom we exchange. We cannot know, or accurately imagine, if another feels benevolently towards us unless we are familiar with them, but we can far more accurately suppose that they will act out of reference to their own interests. Once again self-interested trade and appeals to self-love are the most efficient medium for economic action.

Smith argues that there are two subjective senses of value in the economic exchange of goods. Goods have a value in use, we value a good for its utility to us, and they have a value in exchange, we value a good for what we can swap it for (WN: 44). The complexity of determining value in exchange between differing goods, though it may be achieved by barter, renders trade unwieldy. Questions over how many bags of corn a sheep is worth, or how many sheep a cow is worth slow down and complicate the process of exchange. The acceptance and common valuation of money arises spontaneously as the result of the desire to ease exchange (WN: 284). Rather than exchange raw goods, and face the prospect of not being able to reach a mutually advantageous bargain, humans come to accept some conventional token of common value: Money (WN: 37–46). This allows them to exchange their goods for their 'price' and then exchange through the medium of money for goods produced by others. Money becomes 'the great wheel of circulation, the great instrument of commerce' (WN: 291). It comes to be an indication of the value of goods, having a subjective value of its own in comparison to the amount of those goods that it can buy. For this reason, because money only has comparative value rather than any intrinsic or objective value of its own, it is not the object of human activity. It is sought not as the end of activity but rather as a means to the attainment of those goods for which it can be exchanged: it is a medium of exchange, an indicator of value and wealth rather than value and wealth itself (LJP: 370, 384).[33]

Thus money is an instrument which eases exchange by simplifying calculations of subjective value.[34] The price of a good, expressed in monetary terms, arises from interaction through trade. As Smith puts it: 'It is adjusted, however, not by any accurate measure, but by the higgling and bargaining of the market' (WN: 49). He goes on to point out that there are two types, or senses, of price. There is the natural price, what a good costs to make, and there is the market price, what a good can be exchanged for.[35] This division, analogous to that between value in use and value in exchange (LJP: 358), shows that price in exchange – market price – need not be the same as the natural or 'real', if you like, cost of producing the good (WN: 73; LJP: 361). The market price of a good is determined by the interaction of supply and demand (WN: 76), and is in this sense subjective – that is dependent on the circumstances of the particular exchange (WN: 57). Market price tells us as much about the conditions of exchange as it does about the value of a good: the market serves as not only an arena for the exchange of goods, but also as an arena for the exchange of information about those goods and their production and retail. The key factor in this process is the market price, an indicator which may be read by individuals and which guides their actions.

> The market price [Smith writes] of every particular commodity is regulated by the proportion between the quantity which is actually brought to market, and the demand of those who are willing to pay the natural price of the commodity, or the whole value of the rent, labour and profit, which must be paid in order to bring it thither.
>
> (WN: 73)

The market price of a good or service depends not solely on the cost involved in its production, but also in a large measure on the 'effectual demand' (WN: 73), the demand of those who are in a position to act on their desire for the good. The market price is determined by the interaction of consumer, producer and retailer. This interaction produces, in Smith's view, a spontaneous order or equilibrium which becomes the price of the good and which embodies information about those involved in the transaction and the circumstances under which the good was produced and exchanged. This reliance on circumstances implies that market prices are strongly affected by 'accidents' (WN: 78), particular circumstances of say, geography or weather, which alter the value of the good.

Prices change because producers and consumers continue to act from the same motives (that is to say in pursuit of their own interests) as the circumstances around them change. If a good becomes scarce a consumer who still wants that good and is able to offer more will do so in order to secure the good. The producer, acting on the information they receive from this will then alter their price accordingly. As we discussed before, the 'mercenary exchange of good offices' (TMS: 86), though only one feature of human motivation and interaction, is the force that drives such commercial activity.[36] If self-interest is the incentive to production, and trade operates through the medium of appeals to self-interest, then the motive that drives this process, the desire to fulfil our desires, plays a key role in the determination of the market price of a good. It is this self-interest that keeps prices accurate. As a price indicates the interaction of supply and demand so it is itself determined by the motives of suppliers and consumers and its accuracy is ensured by the regard of each for their own best interest.[37] The desire to profit from our labour is the incentive that prompts us to work. Gain, or the satisfaction of our natural urge to improve our position is the prime motivation of economic activity. And as specialization advances, and the market expands, producers come into competition with each other. The desire to profit, to secure business, leads to an improvement in the provision of goods as well as of the goods themselves (EMPL: 302). Productive techniques are honed to ensure success, the goods themselves are improved to attract customers, and the competition between suppliers drives the price down as they each seek to undercut their rivals (WN: 595).

For Ferguson commerce becomes an object of study (ECS: 58). The individual's desire to fulfil their self-interest in competition with other producers adduces them to apply themselves to the improvement of their skills, their production and the product of their labour. The drive of self-interest prompts technological progress and the refinement of productive methods: in short it promotes an increase in human knowledge. For this reason the key to success in economic activity is to know yourself and your situation and to act accordingly, or prudently. Moreover, and related to our earlier epistemological points, each individual is best placed to exploit their own situation efficiently. Price, in addition to carrying information about supply and demand and in being kept accurate by self-interest and a desire to improve our position, also carries information about the concrete circumstances of individuals. As Smith describes it:

> the private interests and passions of individuals naturally dispose them to turn their stock towards the employments which in ordinary cases are most advantageous to the society. But if from this natural preference they should turn too much of it towards those employments, the fall of profit in them and the rise of it in all others immediately dispose them to alter this faulty distribution. Without any intervention of law, therefore, the private interests and passions of men naturally lead them to divide and distribute the stock of every society, among all the different

employments carried on in it, as nearly as possible in the proportion which is most agreeable to the interest of the whole society.

(WN: 630)

Prices are information signals rendered in a monetary form that assist in the co-ordination of economic activity. They provide a method of 'rating or estimating' (EMPL: 285) labour and commodities that may be read by the various parties and used as information upon which to base their decisions regarding the good in question. Smith indicates this when he shows how the focus of labour is guided by the price of a good, and how the notion of comparative advantage is indicated through the medium of prices. He writes: 'It is the maxim of every prudent master of a family, never to attempt to make at home what it will cost him more to make than to buy. The taylor does not attempt to make his own shoes, but buys them of the shoemaker. The shoemaker does not attempt to make his own cloaths, but employs a taylor' (WN: 456–7). The actions of producers and consumers are guided by the information held in prices. The price of a good carries information about the profitability of a particular occupation and guides people in their decisions as to their choice of specialization. The price of labour (WN: 103) being in a high degree influenced by the level of skill attained, by the human capital, and also by the supply of suitable labourers, implies that wages become similar information signals. They fulfil the same information-exchanging role and are guided by the same self-interest to reflect accurately the conditions of the industry in question. The argument is that prices are used by consumers and producers to co-ordinate their activities. The efficient operation of this price mechanism allows individuals to adjust their behaviour to the concrete circumstances of supply and demand. The central point here is that this informational role of prices is not intended or consciously created or undertaken: we do not intend to send signals by our production and consumption, all that is desired is the satisfaction of our own wants and needs.

Now this, as we know, occurs in a system where, owing to specialization, consumers and producers are ignorant of the details of each other's situation with their attention being focused on their own narrow field. So the price mechanism acts as a simplifying device that allows an individual to process the implications of knowledge which they cannot profitably possess if they are to concentrate on their own specialized occupation. This is achieved at the same time as the simplification of trade as a whole. And the result is that monetary pricing acts as a simplifying medium between the inhabitants of a complex and interdependent economy. Indeed the manipulation and study of prices comes to be of such importance in developing commercial societies that a distinct profession of individuals arises whose occupation and livelihood depend upon the successful reading of price signals. This group of merchants develop skills related to the reading and processing of price signals and the information that they contain about supply and demand (WN: 530). The activity of merchants as specialists facilitates trade by removing the bargaining process between individual consumers and producers. A producer who sells to competing merchants limits the number of individuals with whom he must trade allowing more simple flows of information and simplified interaction. As Hume puts it: 'Merchants . . . beget industry, by serving as canals to convey it through every corner of the state' (EMPL: 301). Competition between merchants adds a further level of efficiency to the process of exchange, benefiting consumers by the manipulation of profit margins that characterizes such competition (WN: 669).

The efficiency generated by competition allows Smith to argue that commerce operates most efficiently when left free from restriction, giving competition rein to govern price (WN: 116). We have already seen that the Scots link the growth of freedom with their notion of progress (ECS: 203), but here we see that freedom, in the sense of free trade, is also related to progress by the efficient functioning of competition and prices. Freedom is both enhanced in the process of progress and vital for the continuation of the process.[38] Freedom arises gradually and through a process of unintended consequences, and as we saw in the Scots' analysis of the decline of slavery and feudalism and the development of the division of labour, progress towards a commercial society is related to the extension of liberty throughout a society. The function of a commercial society, of a system of 'natural liberty' and free trade, is to increase wealth (WN: 324, 372). Freedom of trade acts to do this by allowing the efficient functioning and reading of price signals leading to the efficient exploitation of comparative advantage (WN: 533). Smith highlights this in his argument against restraints on trade aimed at supporting certain home markets; he writes:

> By means of glasses, hotbeds, and hotwalls, very good grapes can be raised in Scotland, and very good wine too can be made of them at about thirty times the expence for which at least equally good can be brought from foreign countries. Would it be a reasonable law to prohibit the importation of all foreign wines, merely to encourage the making of claret and burgundy in Scotland? But if there would be a manifest absurdity in turning towards any employment thirty times more of the capital and industry of the country, than would be necessary to purchase from foreign countries an equal quantity of the commodities wanted, there must be an absurdity, though not altogether so glaring, yet exactly of the same kind, in turning towards any such employment a thirtieth, or even a three hundredth part more of either.
>
> (WN: 458)[39]

We have already seen that the decline of feudalism and the advance of the division of labour leads to a diffusion of wealth throughout society, and that interdependence allows increased and improving material production leading to a situation where even the simplest worker in a commercial society enjoys a level of material comfort beyond that experienced by the wealthy in previous times. The interdependence that produces the labourer's woollen coat may provide better for their needs than an African chief, but it does not provide for their needs in an equal measure to others within their own commercial society. Smith notes that the commercial system and the division of labour are based on a prevalence of inequality in society. The worker is better provided for materially and in terms of freedom than they were in prior ages (WN: 420), but this does not result in any greater sense of material equality. There is a gradual trickle-down of wealth and freedom through the process of historical progress (LJP: 566).[40] The pursuit of luxury goods by the feudal lords encourages the practice of commerce while simultaneously freeing the peasants from dependency and opening the route to prosperity to them.[41] In this sense, as national wealth is increased by the development of commerce, so too, as a result of the invisible hand, is this wealth diffused through society. Wealth diffuses and the value of wages rises (WN: 96), but inequality remains. Indeed, for a commercial society to operate efficiently it must remain. The rich, in general terms, remain rich, but this is not a problem for the Scots' analysis of commercial society because commerce also enriches the poor. The advantages once enjoyed as luxuries by the wealthy are gradually made available to the whole of society (WN: 260). Smith offers a clear example of this when he notes that what was once the seat of the Seymour family is now 'an inn upon the Bath road' and the marriage-bed James the VI and I ended up as the 'ornament of an alehouse at Dunfermline' (WN: 347). Material advantages spread gradually through society with progress being characterized not by the sudden acquisition of a product by the whole of society, but by the gentle diffusion of advantages with the passing of time.

Natural liberty operates through self-adjustment, if left alone human interaction efficiently proceeds along the signals offered by prices. Just as a great legislator could not have been responsible for shaping the whole of a society, so a politician cannot hope to direct economic activity as efficiently as a system of natural liberty. Natural liberty operates efficiently precisely because it is not restrained or directed, it is free to react to circumstances and the accuracy of the information that it passes depends on this. The perversion of prices can be harmful because it imbalances the information which they conduct and affects man's ability to make informed judgements from prices (WN: 632). A number of forces can pervert prices but chief amongst these is the activities of government. As Smith writes of his system of natural liberty:

> The sovereign is completely discharged from a duty, in the attempting to perform which he must always be exposed to innumerable delusions, and for the proper performance of which no human wisdom or knowledge could ever be sufficient; the duty of superintending the industry of private people, and of directing it towards the employment most suitable to the interest of the society.

(WN: 687)

Perhaps the most obvious perversion of the price mechanism arises from the phenomenon of monopoly.[42] A monopoly for Smith is a combination of individuals that seeks to pervert the price mechanism to their own advantage by controlling the supply of a good. Monopoly is, for Smith, a bad thing. He believes that it exists as the result of a particular relationship which develops between merchants and the government (LJP: 527; WN: 452, 613), that is to say that some group of merchants is able to persuade the government that it is in the national interest that they be given special support, usually in the form of restricting entry to their markets, which will place them at an advantage. Smith argues that this advantage for the merchants is bought at the expense of the consumer (WN: 617) and of other producers (WN: 662). Moreover, he is clear that it is the direct result of government action (WN: 174). Monopolies work by raising prices to 'unnatural' levels (LJP: 363). The policies that guide them operate by focusing on production alone rather than on production and consumption. Smith believes that such a situation is a perversion of the interaction of supply and demand; that focusing on the balance of trade from the point of view of producers perverts the efficient operation of the market (WN: 488–9). This is a result of the effect of monopolistic or mercantilistic restrictions on the information carried by prices. As he puts it:

> No regulation of commerce can increase the quantity of industry in any society beyond what its capital can maintain. It can only divert a part of it into a direction into which it might not otherwise have gone; and it is by no means certain that this artificial direction is likely to be more advantageous to the society than that into which it would have gone of its own accord.
>
> (WN: 453)

Restrictions on trade misdirect capital and labour by perverting the accuracy of prices. Perhaps this is most apparent in the area of international comparative advantage. Here the price of a home-produced product is compared with that of importing the same product. If the import is cheaper, in Smith's view, then any restriction which discourages importation of that good misdirects the flow of capital and labour within the country and creates an inefficient industry. Monopolistic and mercantilistic restrictions lead to inefficiencies in trade by restricting the 'free concurrence' (LJP: 364) of prices that provides accurate information. As we noted above, Smith links such practices to the actions of government; he argues that: 'Such enhancements of the market price may last as long as the regulations of police which give occasion to them' (WN: 79).[43] A monopoly of the mercantilist sort cannot subsist without the connivance of a government. Smith believes that merchants are able to secure this support because they are able to persuade governments that what is in the interests of the merchants is the same as the national interest (WN: 475). That this is possible is because the merchants' knowledge of their field and their understanding of the origins of their profits is superior to that of any government official (WN: 434). By appealing to an identification of their own interests with those of the nation, merchants were able to exploit the prestige of their supposed specialist knowledge of trade to extract concessions which were indeed in their interest but which bore little relation to the national interest as a whole. For this reason Smith is wary about trusting businessmen in matters of policy (WN: 471), for it becomes clear that their interests as a class are never wholly at one with the best economic interests of the nation (WN: 145).

Another area of the perversion of prices by group interests is the case of guilds. Guilds operate by controlling a monopoly of the practice of a particular trade (LJP: 84). By their control of skilled labour they are able to pervert the price of labour by limiting access to the profession. As a result they work by restricting trade through the restriction of access to the skills necessary to that particular trade, or in other words, they control access to experiential or non-verbalized knowledge through restrictions on numbers and apprenticeship schemes (WN: 143). The reason traditionally advanced for these restrictions is analogous to the arguments of those merchants seeking monopolistic privileges: that they are best placed to understand and control the interests of the nation in a particular trade. By enforcing professional standards through apprenticeships the nation benefits by the increased skill of the guild-approved tradesmen. Smith, however, rejects this view, arguing instead that: 'The real and effectual discipline' (WN: 146) that ensures competence in workmen is the fear of losing employment.[44] Smith is dismissive of the claims of unions and local corporations to provide increased levels of professional knowledge (WN: 144–5), arguing instead that their chief purpose is to pervert prices to the advantage of their members. However, behind such restrictive practices always lurks the complicity of a government who, by giving legal force or recognition to such monopolistic practices, allows the monopolists to perpetuate them by rendering them immune from competition.[45]

The division of labour and public goods

We have dealt thus far with those aspects of government activity that the Scots considered as necessary to the continued effectiveness of a commercial society. There are, however, other areas in which the Scots believed that government action was required in a commercial age. Ferguson notes that: 'The boasted refinements, then, of the polished age, are not divested of danger. They open a door, perhaps, to disaster, as wide and accessible as any of those they have shut' (ECS: 219).[46] Problems, malign unintended consequences, arise from the process of the division of labour and the division of knowledge; and these problems threaten to undercut the process itself by destabilizing society. The division of knowledge leads, as we have shown, to a fragmentation of knowledge. Specialization necessarily restricts the attention of workers to one particular field and this field, in the case of many workers, will be a simple operation requiring little thought for its exercise. Smith waxes eloquent on the danger of this phenomenon:

> The man whose whole life is spent in performing a few simple opera-
> tions ... has no occasion to exert his understanding, or to exercise his
> invention in finding out expedients for removing difficulties which
> never occur. He naturally loses, therefore, the habit of such exertion, and
> generally becomes as stupid and ignorant as it is possible for a human
> creature to become. The torpor of his mind renders him, not only inca-
> pable of relishing or bearing a part in any rational conversation, but of
> conceiving any generous, noble, or tender sentiment, and consequently
> of forming any just judgement concerning many even of the ordinary
> duties of private life.
>
> (WN: 782)

There is a very real danger that, as the cumulative sum of human knowledge advances by specialization, the individual sums of knowledge (or the scope of those sums) of a large part of the population may fall to levels below that which they would hold in a less developed society. Smith advances a possible cure for this apparently necessary evil of the process of specialization: a cure that is to be found in yet another division of labour and species of special-ization. That is the creation of a specialist group of professional teachers whose job it is to provide a universal system of education (WN: 786). Edu-cation becomes a method of enlightenment and social control, preventing the possibility of disputes that may arise from the susceptibility of a dead-ened workforce to the forces of religious enthusiasm, by socializing them and providing them with a degree of understanding that they would not gain from their everyday employment.[47] Education and the growth of leisure industries provide an outlet for individuals in those specializations that dis-courage extensive thought (WN: 796). They also bring them into contact with others and preserve the process of socialization through mutual sym-pathy. Education also has the advantage of increasing the knowledge of indi-viduals, which in turn contributes to the cumulative sum of human knowledge and encourages the possibility of innovation. Moreover, the divi-sion of labour that creates a leisure industry opens up a new area of commer-cial activity and a new market which offers the possibility of employment and profit. Smith also describes in detail the nature of his proposed educa-tion system, arguing that the levels of education ought to cater to the intended career of the individual allowing them the opportunity to acquire a level of skill that might prove useful to them. Smith's system of education is to be subsidized by the government: he argues in favour of private teachers whose wages are paid partly by the government and partly by the parents of the pupils in an attempt to ensure the provision of incentives which encour-age effective education.[48]

Ferguson, however, is not so sure that education and entertainment are sufficient to counteract the possible ill-effects of the division of labour. He believes that the problem is not so much one of ignorance, but rather one of self-interest detracting from individuals' ability to act in the public sphere (ECS: 177–8). His chief preoccupation in this matter is the effect of the division of labour on military forces, in particular the famous 'militia question' of the Scottish Enlightenment.[49] Millar notes that the spread of commerce and the diffusion of wealth lead individuals to be less willing to enter into military service on behalf of their country. As a result the institution of standing armies funded by taxation replaces the ancient practice of citizen militias (Millar 1990: 222). Ferguson is convinced that this is a dangerous development. He argues at length for the superiority of citizen militias, making frequent reference to classical precedent and to the potential danger to the state of a standing army.[50] His admiration for the citizen armies of Athens and Sparta is qualified by the realization, as Hume and Smith note, that such institutions were only possible because of the prevalence of slavery in the ancient world.[51] Nonetheless Ferguson argues that the development of standing armies is dangerous to the political stability of a nation.[52] Such armies become dangerous as they are open to alignment with political factions and can become a force in internal politics that threatens the stability of government (ECS: 256).

Smith, however, is more sanguine. While deploring the advent of cowardice among a people (WN: 787) he points out that standing armies are undoubtedly more effective for national defence in a modern commercial age (LJP: 541).[53] Smith's argument is a detailed study of the military viewed through his four stages schema. He argues that standing armies are a necessary development as they reflect the unfolding of the division of labour. As technology and skill advanced, the military became a distinct profession practised by specialists who made use of their specialized knowledge. The advance of technology introduces the division of labour within the military profession: artillery and modern weapons lead to the development of distinct sub-disciplines within the armed forces (WN: 689–708). Where Smith does approve of the militia it is not from any belief in the military effectiveness or superiority of citizen troops, but rather it is from the encouragement which participation affords to the fostering of social cohesion and a sense of civic virtue (WN: 787).

Ferguson's concerns, however, about the 'dismemberment' (ECS: 218) of the human character as a result of specialization also lie in this area. He believes that the self-interested pursuit of private gain distracts citizens from the serious business of the public good (ECS: 212). Action in the pursuit of the public good is, in Ferguson's view, an essential part of the human character (Ferguson 1994: 290–1). Self-interest in wealth accumulation renders individuals unwilling to expend their attention on matters of public concern (ECS: 213). Just as a standing army proves dangerous to a state, and the loss of martial skill leaves it open to attack, there begins to develop a political division of labour which is equally threatening to the stability of the state.[54]

The separation of political and military skills into distinct professions, in Ferguson's view, damages social cohesion and goes against human character. It tends to break the 'bands of society' (ECS: 207), and leaves the care of the public good ill provided for as each individual immerses themselves in their own concerns.[55] No one is left qualified to act in the public interest. It is for this reason that Ferguson argues for the restriction of the division of labour in military and political matters. Smith, however, does not go along with this explicitly civic republican view. He argues instead that the rise of a profession of specialist politicians can in fact be a positive development. The increasing complexity of a commercial society, in Smith's view, positively requires the division of labour in government. Smith argues that the division of labour in the departments of government and branches of the justiciary are the product of an unconscious reaction to the circumstances of an increasingly complex society (LRBL: 176). That is to say departments of government and legal institutions such as minor magistrates and juries are introduced to ease the work of law-making and to reap the benefits of the utilization of specialist knowledge (LJP: 88, 283). To ease the workload of a superior magistrate there develops a gradual delegation of power that diffuses responsibility and power throughout the legal and political system in reaction to increased complexity and workload. A further unintended consequence of which is the diffusion of power through society in a process that enhances freedom.

Ferguson's concerns are the concerns of an inhabitant of a classical republic, but he was not living in or writing about small city-state republics.[56] Rather he wrote at a time when a nascent commercial society was developing: a society whose complexity and reliance on specialization demanded a form of representative rather than direct democracy. Hume and Smith are particularly quiet on the republican concerns that moved Ferguson. This, perhaps, was because of their realization that they were indeed experiencing a new form of society where specialization was essential to continued progress. But it is more likely that their focus on the social generation of values through socialization and sympathy as a central aspect of human society, led them to believe that the natural sociability of humans and, indeed, human nature, would prevent the bands of society from being broken in an age of increasing interdependence.

We have seen that the Scots' analysis of the origins and internal operations of a commercial society is conducted through their spontaneous order approach. The division of labour evolves from the unintended consequences of particular self-interested actions to produce a system of interdependency that allows the exploitation of complex specialist knowledge. The benign spontaneous order that is produced as a result of this process is explained by an invisible hand argument. This argument refers to the particular combination of evolved institutions – the rule of law and money –

and practices – self-interested trade and competition – that allows the efficient exploitation and co-ordination of specialist knowledge. The conclusion that the Scots draw from this descriptive argument is that free trade, undertaken within the evolved institutions of the invisible hand, produces socially beneficial results.

6   The evolution of science

Having completed our analysis of the Scots' approach to the explanation of science, morality, law and government and the market, we now move on to examine the same topics in the writings of the theorists of the twentieth-century classical liberal revival. As we move through our analysis we will see how this group of thinkers develop the key concepts that we identified as typifying the spontaneous order approach in the work of Smith and the Scots. The central figure in our discussion, and the foremost exponent of the idea in recent times, will be F.A. Hayek. And by examining his work, and that of his fellow liberals, we will further develop our composite model of the spontaneous order approach.

The impetus to science

When Hayek comes to consider the philosophy of science it is clear that he agrees with Adam Smith's analysis of wonder as the root of the human desire to practise science. Hayek argues: 'Man has been impelled to scientific inquiry by wonder and by need. Of these wonder has been incomparably more fertile. There are good reasons for this. Where we wonder we have already a question to ask' (Hayek 1967: 22).[1] Hayek argues that the recognition of a regularity, or pattern, leads us to pose the question as to why and how this arises. A newly experienced pattern or recurrence of events surprises us, piques our curiosity, and leads us to enquire after the principles behind it. We seek to understand such patterns in terms of some common feature or regularity of circumstance that links the occurrences. In brief we seek understanding to stabilize our expectations and to satisfy our curiosity.

Karl Popper agrees with Hayek on this point, noting that the impetus to science is the desire of 'satisfying our curiosity by explaining things' (Popper 1972: 263). Popper, however, develops a far more complex and nuanced analysis of this situation: one which he deploys throughout his philosophy of science. Beginning from the assertion that our responses to the environment are the basis of enquiry, and grounded in the importance and centrality of wonder, he develops a detailed critique of notions of induction. Popper argues that science does not begin from the position of conscious observation of a phenomenon, rather it is prompted by a problem-situation which arises from practice or everyday experience. His point in making this assertion is to demonstrate that scientific enquiry does not start from the conscious observation of data, but rather arises from the arousal of our interest in that data. What he means by this is that science cannot simply be observation – for what are we to observe? – but is instead 'focused observation': examinations prompted by and focused upon a particular problem (Popper 1989: 46).

Science is not the collection of observed data; it is in reality the collection of theories about phenomena. It deals with relationships between phenomena in such a way that our understanding is always comparative. Such comparisons give rise to classifications of like events in a process that mirrors the discernment of regularities which is constitutive of the human mind. These classes form the basis of the problem situations with which science is a conscious attempt to deal. As Hayek would have it:

> Science consists . . . in a constant search for new classes, for 'constructs' which are so defined that general propositions about the behaviour of their elements are universally and necessarily true. For this purpose these classes cannot be defined in terms of sensory properties of the particular individual events perceived by the individual person; they must be defined in terms of their relations to other individual events.
>
> (Hayek 1976: 174)

All perception, and all science, is based on a process of comparison and classification, which explains what is new in terms of its relationship to what is familiar. Such classifications are mental conceptions that bear no physical relation to the phenomena observed, but rather reflect the ordering process of our own consciousness. Hayek believed that the process of scientific classification is a conscious rendering of an already extant subconscious process of classification that typifies the operation of the human intellect and human perception (Hayek 1976: 108; 1978: 38). The mind itself is defined as an order of classification, a regularity of neural impulses affected by discerned regularities between external phenomena (Hayek 1976: 16). All classification is the manifestation of a human propensity to order that which is experienced. One implication of this is that the classificatory structure of the mind exists as a series of higher order rules, by which Hayek means that consciousness is necessarily dependent on the non-conscious pursuit of the ordering process of the mind. Popper follows a similar line of argument when he notes a human psychological 'need for regularity' (Popper 1972: 23). He views the 'propensity' to search for regularities as a key feature of the human mind. However, Popper stresses the point that this process of ordering is a mental phenomenon: we do not passively wait for an order to become apparent to our minds, but instead actively seek to order that which we experience. For Popper this active ordering is a product of the desire to dispel wonder and to stabilize expectations (Popper 1972: 24).

One feature of this approach of Hayek and Popper is that it leads them to stress the point that such mental classifications are necessarily abstract in that they reflect the mind's 'construction' of classes of phenomena rather than any essential physical similarity of those phenomena. The order that arises from such classifications is a mental phenomenon, based around our understanding of our own perceptions rather than anything which exists essentially in the phenomena so ordered. It is the process of simplification in the face of diversity, the process of discerning similarities and regularities in the external world. If mental classification is based on theorized similarities about experience of the external world, this suggests the view that the ordering of the mind is a relational or comparative order. We classify phenomena by subjective comparison in a process that presupposes the possibility of similarity through the selection of shared characteristics (Hayek 1978: 72; 1979: 48). Hayek refers to this process of classification as the creation of mental 'maps' or 'models' (Hayek 1976: 115, 179) which constitute human understanding. Scientific inquiry is the conscious pursuit of this process of classification or, to be more accurate, the pursuit of the refinement of the classifications that form the order of the human mind.[2] Popper defines the aim of science as being the provision of 'satisfactory explanations' (Popper 1972: 191), explanations that dispel our sense of wonder by presenting a theory which is supported by factual observations. The improvement or advance of science is the development of increasingly satisfactory theories that dispel doubts, or fill 'gaps', in previous theories.

Though science is a formalized rendering of the ordering process of the human mind it should be remembered that the original mental ordering takes place on a subconscious level. It is not deliberately undertaken: rather ordering is carried on subconsciously by all humans as it typifies the very nature of our mental processes. This is highlighted by the significance of our habit of responding in a similar manner to like phenomena. Habit, in this sense, is an essential and subconscious manifestation of the very nature of the working of the human mind. Hayek adds that he believes that just as habits must be acquired through repetition, so the content of the classificatory order of the mind is acquired by exposure to perceived regularities of action or the repetition of phenomena which the mind has ordered, through the process of comparison, as being similar (Hayek 1976: 47). This in itself implies that the retention of experience, memory, is itself a process akin to habit. That is to say it is an 'expression' of the observation of repeated regularities of phenomena (Hayek 1976: 136). Our entire mental process of classification is based on a discernment, or more accurately an imposition, of regularity: a habitual acceptance of the similarity of certain phenomena based on the observation of perceived common characteristics in line with principles developed in our minds (Hayek 1976: 121).

Hayek's analysis of the operation of habit is not wholly shared by Popper.

Popper rejects Hume's causal model with its reliance on habit and repetition (Popper 1989: 42–4; 1972: 7). He argues at length against Hume's solution to the so-called problem of induction, noting that Hume's belief that induction was invalid, though nonetheless was how the mind worked, grounded science on an unnecessary assertion of the power of belief and habit (Popper 1972: 100). He rejects Hume's psychological explanation of causation and links his argument into a general sense of misgiving about classical empiricism (Popper 1989: 23). Popper argues that such empiricism, the belief in the key role of observation, opens reason up to problems of infinite regress and, more importantly, that it fails to account for the fact that it is possible to construct a theory about a phenomenon without observing it (Popper 1989: 138). Moreover, and related to his argument about wonder, Popper argues that if repetition or similarity is the basis of causal assertions then this presupposes the existence or development of a theory of similarity, which in itself is a conscious act of theorizing about that which is observed (Popper 1972: 24).

He goes on to develop his own approach to the philosophy of science in an attempt to solve such problems of classical empiricism. As we noted above Popper argues that in order for observation to take place effectively it is necessary for the attention to be focused by the consideration of a problem or question. He asserts that humans react to such problems by conceiving prescriptive theories that they then examine through observation. Human beings are always theorizing, and, as the mind is a process of classification, all human understanding is posited upon a subconscious process of theorizing (Popper 1989: 220).[3] What Popper notes is that classical empiricism is flawed because of its mistaken belief that theories are drawn from observation. In his view the reverse is the case, observations are made to test theories. In scientific terms these theories are 'tentative hypotheses' (Popper 1961: 87) that are submitted to the test of the observation of the factual circumstances which they purport to explain. Observation and experimentation are used to eliminate those theories, or aspects of the theory, which fail to conform to the evidence: that fail to explain what they claim to explain. This leads Popper to his famous assertion that the true scientific nature of a hypothesis is not that it is open to absolute verification, but rather that it is open to falsification by a process of empirical observation. Attempts to falsify a theory allow the identification of weak points or 'gaps' in its formulation thence allowing us to 'weed out' (Popper 1961: 133) unsuccessful theories and enquire after new, more satisfactory, explanations. One such empirical test is, of course, that of repetition or constant conjunction, which, rather than being the source of our habitual belief in causal links, is instead a criterion against which to measure our tentative hypotheses and theories about causal links (Popper 1989: 53).

Popper illustrates his theory of the advance of scientific understanding in precisely the same manner as that deployed by Adam Smith, by demonstrating its unfolding in the history of astronomy. Popper argues, as Smith did, that all theories are tentative hypotheses that may be overthrown (Popper 1972: 29). As a result there is no guarantee that because a theory has survived empirical testing in the past that this will continue to be the case in the future (Popper 1972: 69). In this sense a theory can be regarded as positive, or the best which we at present have, solely on the grounds that it has survived testing thus far and not, on any account, because it represents absolute truth (Popper 1972: 15, 20). Popper illustrates this by arguing that neither Newton nor Einstein believed that their theories were the 'last word' (Popper 1972: 57) or represented absolute truth. Rather they both worked on the assumption that they were engaged in a process of immanent criticism, reacting to 'gaps' or problem situations in existing theories that had become over-stretched or dissatisfactory in terms of explanatory power. Popper argues that new hypotheses, if they are to fulfil the role of plugging a 'gap', must both succeed where the previous theory succeeded and surpass it by filling the 'gap'. He illustrates this by noting that Newton's theory succeeds Kepler and Galileo, but also manages to contain them (Popper 1972: 16). It succeeds where they succeeded and also in areas where they left 'gaps'. Newton's theory unifies Kepler and Galileo (Popper 1972: 197). But this unification is not achieved by a sleight of hand or an innovation within the existing theoretical framework. Newton did not deduce from past theories, instead he reformulated the problem situation (Popper 1972: 198) and developed a new hypothesis that took cognizance of the 'gaps' in past attempts. The process is not one of a gradual aggregation of theories, of a collective bundle of observed or tried and tested approaches which must be absorbed, but rather is a process of theories being supplanted or overthrown by new hypotheses which are equally open to falsification.

Hayek broadly shares Popper's views on these matters. Science is the process of plugging 'gaps' in existing knowledge (Hayek 1967: 17). But of course, following Popper, Hayek is keenly aware that such a process of theorizing is grounded on the formation of hypotheses which are open to falsification through experimentation (Hayek 1979: 29; 1967: 28, 32). The test of a hypothesis is if it holds in experiment, if it is consistent and not contradicted. In this sense the strength of a scientific assertion is not the evidence which supports it so much as its openness to refutation by future experimentation, what Popper calls critical rationalism. Such an approach suggests that the advance of human knowledge is often about eliminating what is false and thereby moving closer to truth. Explanation is a negative process of trial and error, one of weeding out false or ill-conceived assertions and hypotheses once the 'gaps' in them have become apparent through observation (Hayek 1979: 74; Popper 1972: 74).[4] As Hayek, following Popper, puts it: 'science does not explain the unknown by the known, as is commonly believed, but, on the contrary, the known by the unknown' (Hayek 1967: 5). This is not, however, to say that Hayek and Popper have forgotten about the role of science in stabilizing expectations. Far from it, the process of hypothetico-deductive falsification is a process of whittling down or lessening the variables in a given situation. The elimination of false, or flawed, assertions narrows the scope for future error (LLL vol. 2: 54); it stabilizes our expectations and it brings us closer to the truth by reducing the chance of possible wondrous events.

As a result of his approach to the human mind Hayek follows Hume in his assertion that the science of man is necessarily the basis of all science (Hayek 1979: 40). If all our understanding is based on our mental classification of the external world, then a proper understanding of the human mind is a necessary prerequisite for the further advance of the natural sciences. A further point should be made here, namely, that this process of classification which typifies the human mind is supposed to be universal. Though the modern liberals have grave doubts as to the accuracy of any detailed conception of human nature (Hayek 1960: 86), attributing much of what the Scots believed the concept to include to the influence of traditions of morality, they nonetheless reject the more extreme forms of cultural relativism (Hayek 1991: 120). Hayek in particular appears to operate with a 'pared-down' conception of human nature that posits a series of underlying universal phenomena which apply to all humans. In *The Sensory Order* he expresses the belief in certain universal emotional and 'biogenic' (Hayek 1976: 96–101) needs and drives which can be discerned in all creatures that we recognize as human. Biogenic drives such as hunger and thirst are universally experienced by the corporeal frame and in turn induce certain attitudes in the order of the mind – the desire for food and water. Originating in the physical needs of the body they provide a series of typical sets of attitudes and actions that may be identified as underlying much human behaviour. Thus Hayek argues that humans universally desire to preserve their lives (Hayek 1988: 69), they universally desire and seek food, shelter and sex (Hayek 1967: 314), and such underlying universal characteristics, drawn from biology, affect a series of universal mental attitudes in human psychology.

Beyond such biogenic similarities, the species also expresses a universal underlying similarity in the operation of the human mind.[5] The mind, as we have seen, is a system of classification and categorization and, while the content of such classifications is largely learned from experience and socialization, the fact that this is how the mind operates represents an underlying universal characteristic of what it is to be human. Our understanding of others is based upon our self-understanding, which is to say that the process of the classification of experience that constitutes the human mind recognizes like actions in others. Hayek's example here is that, based on our own experience, we are able to recognize in another a distinction between a conscious action and an unconscious response. Understanding is possible because of an underlying similarity in how the human mind operates (Hayek 1984: 117). We are able to recognize another as being human because our mind classifies them as such when it becomes aware of them as a classifying form of being (Hayek 1980: 63–5).[6] It is this which lies behind Hayek's assertion that humans are 'by nature' rule-following animals (LLL vol. 1:

11). The human mind operates by a classificatory system the contents of which are drawn from experience, and though the particular experiences may differ beyond the biogenic drives, the fact of rule following remains constant. That all individuals' minds proceed by a like process of classification by generalized 'rules' (Hayek 1979: 43) is the underlying assumption that allows human communication (Hayek 1979: 134) and which finds its clearest expression in the development of language (Hayek 1976: 135, 141; 1988: 106), itself a process of classification which expresses mental classes.

Social science

For Hayek, all science is theorizing based around classification and general rules. That this creates a distinctly subjective framework is particularly apparent in the social sciences. Our categorization of social phenomena is what constitutes them. Social phenomena, such as class, nation, government and so on, are mental constructions rather than objective facts or entities (Hayek 1980: 74). As a result, to study a social collectivity, such as class, as though it were an objective entity is a fundamental error (Hayek 1979: 95–7). Instead, Hayek argues, we ought always to proceed in our studies with a clear awareness that such classifications are precisely that: abstractions created by the mental process of classification rather than objective facts. It is for this reason that Hayek proceeds with a methodological individualist approach. Methodological individualism, in its Hayekian form, stresses that in the social sciences we are never dealing with facts as such, but rather with individuals, concepts and opinions. To speak of a social entity as 'acting' is a category error for Hayek, an example of a naive and uncritical anthropomorphism that is at odds with the 'true' understanding of individualism (Hayek 1984: 135).[7] Methodological individualism does not argue that individuals do not associate in social groupings, and it cautions against viewing humans as unattached atoms while at the same time rejecting the view that social groupings are objective facts in any true sense. The approach does not neglect the fact of human interaction, nor does it depend on any psychological theory of self-interest; rather it is the acceptance that the base unit in the consideration of social phenomena, the starting point of enquiry, must be on the level of individuals. The task of social theory, as Popper sees it, is to adopt a methodological individualist approach and use it to build mental 'models' (Popper 1961: 136) or reconstructions of social life in terms of the individuals that compose them.

Following Hume, Hayek argues that social order is based on opinion (LLL vol. 3: 33), and thus social science must be as much an understanding of what individuals believe themselves to be doing, as it should be a study of what they are 'objectively' doing.[8] Hayek also stresses that we are compelled to observe society from the inside; as social beings humans can never remove themselves wholly from a society or its classificatory order if they are to study it. We deploy a sort of parody of Smith's impartial spectator in order to understand others. That is to say that just as communication is made possible by the supposition of a like system of classification of perception, so we, to a great extent, understand the actions of others by understanding our own minds (Hayek 1980: 59–63). This, however, is not to say that the social scientist's object of study is defined wholly by their own subjective opinions, rather it is to assert that the subject matter is the opinion of those whom they study (Hayek 1979: 47). Social science and its conclusions must be grounded in an awareness of the ultimate subjectivism of human understanding (Hayek 1988: 97).

Hayek argues that the confusion of thought which he calls 'scientism' is based on a failure to grasp this subjectivism in the methodology of social science and on a desire to extend the methodological assumptions of natural science into the social sphere. He is keen to stress that the phenomena studied in social matters are necessarily more complex and difficult to pin down than those in the natural sciences. There are for Hayek two distinct types of social science. There are those which are akin to the natural sciences, such as geography and ethnology, and there are those which are more fully focused on the significance of human opinion and motivation and which are, as a result, more fully subjective (Hayek 1979: 42, 67).[9]

The social sciences are necessarily complex in their subject matter, based as they are on theories about complex interactions of human opinion and action. The explanation of complex phenomena such as morality must be based on a study of the individuals who hold such beliefs: individuals who to a large extent may be unaware or incapable of expressing the opinions which guide their actions. As a result, social science must be pursued with an awareness of the peculiar role of individuals who hold opinions and also with the recognition that these individuals possess only partial knowledge of the social processes through which they interact (Hayek 1979: 50; 1980: 54). Hayekian methodological individualism is intimately linked with the concept of dispersed knowledge and unintended consequences.[10] Society and civilization are the result of this process of individual interaction in conditions of limited or partial knowledge. So the subject matter of social science is necessarily the unintended consequences of the interaction of individuals with only limited knowledge of the consequences of their actions (LLL vol. 1: 20). Popper agrees with Hayek on this point, arguing that the proper subject matter of social science are the unintended consequences, which are inescapable, of human action (Popper 1989: 69, 124; 1961: 158; 1972: 117). He believes that it is the task of the social sciences to trace the unintended consequences of human action and to build causal models in order to help explain them.[11]

Before passing on to discuss how Hayek and Popper view the various 'mistaken' approaches to social science, it may be pertinent to pause and say a word about a more explicit example of the compositive method, about how social science research ought to be undertaken. As we noted in Chapter 2, social scientists are faced with the problem of the difficulty of experimentation that the Scots resolve by turning to history as a source of evidence. The problem here, we noted, was the distinct lack of evidence pertaining to the origins of social institutions and practices. This lack of evidence led the Scots to produce their method of conjectural history. Hayek and Popper are also aware of this problem for social theory. They both argue that the Scots were in a great measure correct in their assertion that in the absence of direct evidence it would be useful to construct carefully 'corroborated' (Popper 1972: 189) reasonable models of how social practices might have come about (LLL vol. 3: 156). In the absence of evidence we might attempt rational reconstructions (with no claim to their being actual history) that aim to understand and account for the function of certain social practices in past ages.[12] Popper refers to this approach as 'situational analysis' (Popper 1972: 179) and argues that it is based on a critical reconstruction of 'problem situations' (Popper 1972: 170) in order to provide a hypothetical explanation of the adoption of practices. All such studies will necessarily be conducted with a degree of abstraction and be based on general observations and classifications of the nature of past social practices. They will be, in Popper's terms, 'generalized historical hypotheses' (Popper 1972: 272) which will provide a model of 'meaning' in history (Popper 1966 vol. 2: 278). However, Hayek is keen to caution against the temptation to treat such rational reconstructions as 'facts' (Hayek 1979: 128) or as objective phenomena in a physical sense. Such theories are not, and ought not purport to be, absolute reconstructions of actual events. They should be viewed as the product of situational analysis, or 'compositive social theory' (Hayek 1979: 151) rather than a re-enactment or re-telling of a series of demonstrable events. Conjectural history in this sense is not a universalistic claim about historical laws or necessary processes but is an explanatory model of how historical practices might have arisen which allows us to approach the analysis of their function with an awareness of context and circumstances.

The compositive method of social science that Hayek and Popper identify is predicated on the notion of the centrality of the unintended consequences of human action in the formation of those practices that the social scientist seeks to analyse by the reconstruction of critical models of explanation. The 'anti-historical' (Hayek 1979: 113) and unintended consequence approach of the Scottish conjectural historians to social change militates against the danger of mistaking their conjectural abstractions for facts. The purpose of conjectural history, and indeed of social science as a whole, is to seek the function of social rules and practices which have emerged from the unintended consequences of human interaction; and which are, as a result, to be treated neither as the consciously designed product of historical actors nor as factual wholes which may then be examined and treated as objective entities.

Finally something ought to be said here about the implications of all of these subjectivist approaches. Hayek and Popper are not arguing that in some sense all truth is subjective, they are not in a strict sense relativists (Popper 1966 vol. 2: 261, 269). Popper notes that, in the past, theories that focus on critical judgement had been criticized by some thinkers, such as Bacon and Descartes (Popper 1989: 15), because they viewed it as leading to an unacceptable and isolated subjectivism. The problem of subjectivism mirrors that of taste; knowledge becomes plastic and open to individual interpretation in a manner that appears to preclude any scientific unity or advance of knowledge as a whole. We have already seen how Hayek seeks to avoid this charge by stressing the underlying universality of the structure of the human mind. However, this still leaves open the charge of excessive subjectivity. The question remains how to move from a subjective method-ology to an objective situation that can be the object of debate. Popper's solution to this issue is to be found in his identification of three worlds.[13] He defines these as: 'the physical world "world 1", the world of our conscious experiences "world 2", and the world of the logical *contents* of books, libraries, computer memories, and suchlike "world 3"' (Popper 1972: 74, his italics).

What Popper is stressing here is that objective criticism is only possible of objective assertions. That is to say that once a subjective position is laid down, is published along with arguments in support of it, it becomes an 'exosomatic artifact' (Popper 1972: 286) and we are able to indulge in crit-ical reflection. The written statement becomes the object for discussion by critics and experts who are able to apply rational argument in an attempt to refute the hypothesis. Popper asserts this in order to show that the 'problem situation' of science is in reality the state of the critical debate at any given time (Popper 1972: 107). The 'problem situation' is precisely the discussion of those 'gaps' and problems with hypotheses that have arisen as a result of immanent criticism (Popper 1989: 129). Objective knowledge is the state of the debate.[14] It is the critical discussion of those hypotheses which have thus far survived the process of falsification. For Popper this approach is a key development in the growth of civilization because, put roughly, we have reached a situation where ideas can evolve or die out in the critical discus-sion of written statements rather than through the death of those who hold them (Popper 1972: 66). Objective knowledge, as the state of the debate, is the generation of an objective standard from subjective opinion, it is an interpersonally generated objective value or spontaneous order.[15] As science evolves by seeking conformity with facts, so our knowledge evolves through an interpersonal standard or equilibrium that is the state of the debate between scientists.[16] The attention of scientists is directed by the problem situation which defines current debate, their observations and reasoning have an aim in view: to stabilize expectations and dispel that wonder which makes us curious about gaps in the hypotheses which represent the current level of our knowledge.

A similar argument is advanced by Michael Polanyi, who also notes that the problem situation of science is defined by the state of the critical debate (Polanyi 1969: 50), but he advances the argument a stage further by sug-gesting that the individual scientist's awareness of the state of the critical debate acts as an invisible hand mechanism which allows them to react to the work of other scientists (Polanyi 1969: 51). A scientist's attention is directed to areas of interest by the published work of other scientists and, moreover, by participating in this debate the scientist comes to accept a series of professional standards of inquiry which are necessary to engage others in discussion (Polanyi 1969: 52). These standards of scientific propriety are enforced by the critical discussion of the results of the scientist's work by his peers (Polanyi 1969: 55). From this there emerges not only a consensus as to the state of the debate, but also a consensus as to what is required, in terms of rigour and professional propriety, for a scientist's contribution to be accepted as part of the debate. Scientists become socialized into the tradition of the practice of science, reacting to the criticism and praise of their fellows to produce contributions to the debate (Polanyi 1969: 85).[17] The body of science, for Polanyi, is based on the opinions of the practitioners of science, with the results of a scientist's work altering the content of the tradition but not the overall form of science (Polanyi 1951: 40). Science is at base a discussion of a tradition of thought which is subject to immanent criticism in the sense that the established view is debated and gradually improved upon without the wholesale rejection of the tradition itself. Science and scientific progress are a spontaneous order; they are based on 'twin principles; namely, self co-ordination by mutual adjustment and discipline under mutual authority' (Polanyi 1969: 84).[18]

Scientific hubris

Just as the Scots had taken issue with many of the prevalent approaches to political theory in their own time, so too do the theorists of the classical liberal revival take issue with many of the prevailing methodological assumptions of their immediate predecessors in social thought. They launch a sustained critique of the methodology and fundamental principles which had dominated the social sciences from the nineteenth century onwards. Hayek is clear that the errors of what he refers to as constructivist rationalism stem from a fundamental misunderstanding of the nature of society and the subsequent development of a wholly erroneous approach that, ultimately, misses the point of social phenomena. He argues that the hostility displayed by constructivist rationalists to the operation of social phenomena such as the market is grounded in a naive form of anthropomorphism which, though easily fallen into, is fatal to the study of social institutions.

As we have seen Hayek's theory of the mind conceptualizes human thought as a process of classification grounded on a desire for order and stability of expectation. Natural science represents the ordering through classification of our experience of the natural world, while social science follows a similar process with the vital qualification that in social science our classifications refer not to concrete physical phenomena but rather to mental reconstructions that are necessarily only partial and selective.[19] One result of this is that language has developed in such a way that the words used to connote order usually imply or specify an ordering agent (LLL vol. 1: 26–9). From here it is but an easy step to viewing social phenomena, phenomena produced by humans acting in a social context, as an order produced by those people in a deliberative manner. It follows that the personification of society and social institutions, that is to say the viewing of them as conscious ordering entities whose workings are akin to those of the mental processes of the human mind, becomes an obvious and 'simple' means to make sense of the complexities of human societies. It is this error, the failure of the social scientist to conceive of an order other than in anthropomorphic terms, which leads to many of the failings of constructivist rationalism in Hayek's view. This inability to conceive of a non-anthropomorphic, and yet not strictly 'natural', ordering process is the fundamental error of the great schools of social thought which developed between the Scots and the Moderns.

The eighteenth century's growing admiration for science, the very essence of the term Enlightenment, led to a worship of scientists. This admiration led to a desire to emulate the success of the natural sciences by applying their methods in social studies.[20] According to Hayek this trend led to a misapplication of the legitimate methods of the natural sciences in areas where that approach was not strictly applicable. This is the train of thought which he refers to as 'scientism' (Hayek 1984: 267). The admiration accorded to scientists and 'science' led to the over-eager acceptance of that which gave the appearance of science. In terms of the social sciences it was accepted that the result of 'scientific' inquiries were indisputable. The scientistic mindset leads to the belief that since society is a product of the actions of humans, since it is man-made, it can consciously be remade by humanity to meet predetermined goals (LLL vol. 1: 59). This simplistic view of the nature of social institutions, aside from being factually inaccurate, was also conceptually wrong in its very approach to the study of society. To view social institutions as serving a purpose, in the sense of being designed to fulfil the goals of a 'mind' that operates along the same principles as individual human minds, was the cardinal error of constructivist rationalists.

Hayek applies the generic name 'constructivist rationalism' to those approaches to social science which he believes to be based on a mistaken assumption of anthropomorphism. He defines constructivist rationalism as: 'the innocent sounding formula that, since man has himself created the institutions of society and civilization, he must also be able to alter them at will so as to satisfy his desires or wishes' (Hayek 1978: 3), or put another way: 'a conception which assumes that all social institutions are, and ought to be, the product of deliberate design' (LLL vol. 1: 5). It should be noted that the second formulation introduces something of that supposed superiority of the organized, deliberate ordering which is a feature of the scientistic approach to society. That is to say there is a presumption that what is deliberately designed is superior to what merely 'is', and that if humans do indeed make social institutions, then it is desirable that they should make them in line with a deliberate, rational plan. Hayek believes that this approach is an error because, quite simply, it is factually wrong. It pre-supposes that since humans make social institutions that they do so in a deliberate or conscious manner.

Constructivist rationalism presupposes that social institutions serve some definite purpose, and that they were constructed to serve that purpose. This approach, of course, implies that there was a pre-existing agreement as to the purpose which the institution was to serve. Constructivist rationalist approaches must have an end in view, they must be organized for the attain-ment of a purpose: for otherwise such an institution would be 'irrational' and have no place in a 'planned' society. In this sense constructivist rational-ism contains a high degree of teleological supposition. That is to say, con-structivist rationalists presume not only design, but design for a specific purpose (Hayek 1979: 45). They are constitutionally incapable, as a result of their premises of design and purpose, of accepting the existence and success-ful functioning of institutions that are the product of a process of unin-tended consequences. It is for Hayek the 'hubris of reason' (LLL vol. 1: 33) to judge institutions and spontaneous orders that are the result of a process of cumulative unintended consequences as though they were the product of deliberate design. In other words, the constructivist rationalist's assumption that design and conscious direction are more efficient prejudices them against any form of association not based on these principles. This failure to see the value, or even in some cases the very existence, of non-purposive, undesigned institutions is the root of the failure of their approach.[21]

Hayek traces the anthropomorphic fallacy, which lies at the heart of con-structivist rationalist approaches to social science to two ancient dichotomies set up by the Greeks between 'physei' (by nature) and 'nomo' (by conven-tion) or 'thesei' (by deliberate decision). He argues that these crude distinc-tions, and especially the conflation of the last two, 'nomo' and 'thesei', led to a conceptual misunderstanding which has plagued social thought. It pro-duced a situation where different authors refer to the same social phenomena as natural or artificial; a confusion of terminology that encouraged the iden-tification of the artificial with that which is the product of design.

Hayek credits Mandeville and the Scots with the identification of this dif-ficulty. The success of the Scots, in Hayek's view, was to recognize a third category of phenomena which escaped the Greek conflation of 'nomo' and 'thesei' (Hayek 1967: 97; 1988: 145). This third category, which Hayek defines with a phrase from Ferguson, are those phenomena which are the result of human action, but not the product of human design. Into this cate-gory, Hayek claims, fall the human social institutions that have traditionally and erroneously been approached as being the product of deliberate design. On this way of looking at things the 'social' is necessarily a human product, but not a conscious or deliberately designed product: rather it is the product of a process of unintended consequences (Hayek 1979: 154). The

constructivist rationalist approach represents an alternative to the spontaneous order approach for the understanding and explanation of social phenomena. It is the thrust of Hayek's position that it is deeply flawed as an
approach and particularly inappropriate when applied to social phenomena.

The practical political manifestation of the errors of constructivist rationalism is to be found in the phenomena of socialist planning. Though the
critique of this school of thought will receive more detailed attention in
Chapter 9, it is important to stress here that it is a product of the erroneous
anthropomorphism of constructivist rationalism. Socialists, according to
Hayek, are guilty of a number of factual and conceptual errors arising from
the crude rationalism and oversimplification of constructivist rationalism
(Hayek 1960: 406; 1991: 86). Socialism is a species of collectivism that is,
in essence, an attempt to organize or plan society in such a way as to advance
specific ends of a primarily economically egalitarian nature. Such planning is
posited on the belief that it is possible and desirable to organize national
economies in line with a consciously developed plan to achieve specific goals.
This belief, according to Hayek, is based on what he terms the 'synoptic
delusion ... the fiction that all the relevant facts are known to some one
mind, and that it is possible to construct from this knowledge of the particulars a desirable social order' (LLL vol. 1: 14). The engineering mindset of
constructivist rationalists and socialists leads them to assume that it is possible to centralize all of the knowledge necessary to direct an economy along
a plan in an efficient manner. Here suffice it to say that Hayek believes this
to be a factual impossibility.[22] Hayek advances the view that the attitude to
planning implicit in socialism will lead, inevitably, to greater and greater
intervention and control of economic transactions in order to preserve a
favoured plan, and that such intervention and control will lead to totalitarianism (Hayek 1991: 68). Totalitarianism is the malign unintended consequence of attempts to plan society.

The anthropomorphic misconception that lies behind attempts at social
and economic planning is, for Hayek, the root of the errors of all socialist
movements. Socialism, with its ideal of social justice, is guilty of a fundamental error of categorization. That is to say, the very concept of social
justice is itself an anthropomorphic misconception that attributes the
human value of justice to a phenomenon (society) which is not a human
agent. As we shall see in the following chapters this is precisely the situation
which Hayek believes is impossible: society is not a phenomenon that is
deliberately ordered but, rather, one that is based on certain fundamental
independent self-ordering principles: society is a spontaneous order. The
failure to grasp this 'fact' by those committed to anthropomorphic concepts
such as social justice is a result of a naive personification of society.

A further manifestation of the scientistic approach of constructivist rationalism is to be found in the development of what Hayek and Popper call
historicism. This tradition of thought, on their own particular understanding of it, is produced by an erroneous application of supposedly scientific

methods to the study of history (Popper 1966 vol. 2: 264). Its cardinal error is to treat its theorized concepts as though they were historical facts (Popper 1972: 354). This inclination leads to the belief in 'laws' of historical development which seek not only to order our understanding of history, but to give it meaning (Popper 1966 vol. 2: 278).[23] Historicism proceeds by identifying historical trends and then formulating them into principles that determine historical change and allow historical prophecy. Hayek and Popper both reject the notion of historical laws as being theoretically impossible and factually erroneous for the simple reason that history is adaptation to changing circumstances: it is an evolutionary rather than a teleological or essentialist process. They go to great lengths to dispel the notion that humans are capable of accurate historical prophecy.

They begin their critique by noting the distinct failure of all previous attempts at detailed historical prediction, pointing to the examples of Keynes and, more importantly, Marx (Hayek 1967: 262; Popper 1966 vol. 2: 82); and suggest that such past failures demonstrate the futility of such attempts. Hayek then goes on to attack the notion that scientific prediction is applicable as a method to the study of social theory. He believes it to be a false view that historical science is capable of producing detailed prediction of future events (Hayek 1979: 344). What he argues here is not that social science is incapable of tentative hypotheses about types of possible future events, but that the precise nature of these events is impossible to predict. Popper, however, holds a different view. He argues that the errors of historicism do not reveal that the method of the natural sciences are inapplicable to the social sciences, but rather that historicists have misunderstood the nature of scientific laws. Scientific prediction, Popper argues, is not the same thing as historical prophecy (Popper 1966 vol. 1: 3). Moreover, the more conditional claims of scientific prediction are in no way essentially related to a notion of determinism (Popper 1966 vol. 2: 85). For as we noted before, Popper understands all scientific knowledge as tentative hypotheses open to falsification by future events.

Having dismissed the scientific pretensions of historicists they move on to criticize the actual method applied to discern these predictions. First they note that experimentation is impossible, historical prophecies can only be falsified by events – as Marx's were (Hayek 1979: 73). This introduces what Popper calls the 'Oedipus effect' (Popper 1989: 38) whereby the prophecy itself – say Marx's belief in the inevitability of a revolution – plays an instrumental role in a future event – the Russian Revolution for example. However, as Popper notes, the revolution did not occur for the reasons or in the location which Marx predicted: it was a product of the belief in the prophecy rather than an expression of the truth of Marx's 'laws' of history. Hayek goes on to argue that accurate historical prophecy is a practical impossibility as a result of the complexity of the factors concerned (Hayek 1967: 34). Social phenomena are by their nature enormously complex and interrelated and, as a result, are highly resistant to being reduced to simple laws (Hayek 1967: 20). If the complexity of social events defies prediction by its nature, then the reverse of this assertion is also true. Which is to say that the limited nature of the human mind, the restricted nature of human knowledge, means that no individual or group of individuals could ever command the complete knowledge which would be required to make detailed social predictions (Hayek 1979: 73–4).

However, though detailed prediction is limited by the complexity of the data and the nature of human knowledge, this is not to say that conditional predictions are beyond the social sciences (Hayek 1976: 185). These predictions are necessarily highly tentative and dependent on a long series of absolute qualifications. They might only occur in exactly those situations to which they are designed to apply: which is decidedly not to say that these precise conditions will necessarily occur. Leading on from this Hayek and Popper note a further obstacle to accurate, detailed historical prediction: as we have seen they define the subject matter of social science as being the unintended consequences of human action. From here they argue that it is impossible to foresee all of the consequences of an action that takes place in a complex social environment. Logically this is true by definition, for if one were capable of predicting an unintended consequence, then it would become a part of the reasoning process before an action and in a sense would cease to be unintended (LLL vol. 1: 111).[24] If we are constitutionally incapable of seeing even all the immediate consequences of our own actions (LLL vol. 2: 17), owing to the limited nature of our natural capacities, then we cannot hope ever to be able to discern the more remote consequences of our actions. Accurate predictive foresight in social matters is precluded by the operation of both complexity and the resultant unintended consequences of our actions in a social environment (Popper 1966 vol. 2: 94). Even if it were possible to have an action without unintended consequences, a highly doubtful proposition in a social context, such an action would be of little interest to a social scientist as it would produce no problems for social thought beyond psychology. It could be understood in elementary terms (Popper 1966 vol. 2: 96).[25] If total and accurate historical foresight were possible then, in a sense, there would be no social problems. There would be no need to adapt to unforeseen change because there would be none. Moreover, such a view would compel government action. If the state could foresee the course of history, then surely it would have to act to move events towards that end, once again raising the spectre of the Oedipus effect (Hayek 1991: 57).

Hayek dismisses such approaches and instead argues that human history is retrospectively discernible as the adaptation to unforeseen events and unintended consequences (LLL vol. 1: 54). He argues that it is possible to advance hypotheses about the future so long as they are restricted to assertions about types of phenomena rather than specific events (Hayek 1960: 40; 1967: 13–15, 35; 1984: 256, 328). Hayek cites two examples of this kind of prediction. First he compares it to the formation of a crystal, arguing that science allows us to predict the growth of a crystal in specific conditions but, he notes, it does not provide us with the means to foresee the shape or precise nature of the crystal's appearance (Hayek 1967: 28).[26] He then compares this non-specified prediction to the Darwinian model of evolution. Both Hayek (LLL vol. 1: 23–4) and Popper (1972: 270) view Darwin's theory as a historical rather than a historicist theory. Which is to say that evolution describes a process, but provides no specific predictions as to the precise outcomes. Biological life will evolve, but the precise nature of that evolution is unpredictable. Darwin describes what has happened in the past in evolutionary terms, but cannot give any guide as to what will occur in the future. We could not have foreseen the outcome of the evolutionary process because it is by its nature an adaptation to changing conditions which could not in themselves have been accurately foreseen. Popper, however, stresses that this does not exclude the possibility of the scientific examination of society or history provided that such a study is undertaken in full awareness of the conditional nature of the knowledge acquired.

This leads, however, to the final, and decisive in Popper and Hayek's view, criticism of historical prophecy. Such predictions are impossible because the adaptation to change that is the very stuff of history is one and the same process as the growth of human knowledge. As a result, even if the circumstances could be predicted, the reaction to them, the result of the trial and error process of adaptation, could not. This links back to Popper's assertion that we cannot explain our knowledge as that would require the possession of more knowledge than we had.[27] The same applies to prediction: to predict the future growth of knowledge we would have to have knowledge of knowledge which we have yet to attain (a logical impossibility) (Popper 1961: vi; 1972: 298). In order for such predictions to be possible we would have to end the growth of human knowledge, and even then we would have to know more than that in order to explain it (Hayek 1979: 160).

We ought to mention briefly here that Popper and Hayek link this critique to their instrumental justification of freedom from epistemological efficiency. Intellectual and individual freedom is essential, they argue, if we are to adapt to unforeseen circumstances (LLL vol. 1: 56; 1960: 29). For there to be a chance of accurate historical foresight such freedom for adaptation would have to be restricted to reduce the complexity and impose sufficiently definite conditions to allow prediction (Hayek 1991: 120). However, such freedom for adaptation is vital, especially on the 'boundaries' (Hayek 1960: 394) of our knowledge where we are incapable even of conditional predictions. Theories that claim to be able accurately to predict the future path of human history are to be considered as irrational. This prophetic feature of historicism, the view that social science has as its aim the delineation of laws which allow the accurate prediction of future events is a dangerous error. The modern liberals reject the historicist approach to social theory as a constructivist rationalist error, a misunderstanding of the nature of scientific method that is misapplied to the study of the development of social phenomena. What they go on to elaborate is their own understanding of rationalism, the spontaneous order approach and its application to the study of social phenomena.

Critical rationalism

Having examined the Moderns' critique of constructivist rationalism it remains for us to make explicit the alternative understanding of human reason that they advance: what Popper and Hayek call critical rationalism. Hayek argues that: 'Reason was for the [constructivist] rationalist no longer a capacity to recognize the truth when he found it expressed, but a capacity to arrive at truth by deductive reasoning from explicit premises' (Hayek 1967: 107). This focus on purposive rationality, of using reason as an instrument for the independent deduction of truth, became popular along with the rising admiration for the methods of the natural sciences. It led to an attitude that viewed reason as something objective, something outside humans which could be used by them to deduce universal truths. Such an 'erroneous intellectualism that regards human reason as something standing outside nature and possessed of knowledge and reasoning capacity independent of experience' (Hayek 1960: 24), leads to fundamental errors and a misplaced belief in the effectiveness of reason. The notion that reason exists as a single entity – Reason with a capital R if you like – is simply false in Hayek's view. In his view there is no objective entity of Reason, rather there exist only the limited stocks of reason held by individuals who exist within a social context (Hayek 1984: 136). While constructivist rationalism assumes Reason with a capital R, Hayek instead draws upon a tradition of thought which is often, and he believes wrongly, referred to as anti-rationalist or sceptical. Hayek builds on the approach to human rationality advanced by Bernard Mandeville and the Scots. This critical rationalism entails an acceptance of the fact that human reason is fundamentally limited. That is to say, reason is an attribute possessed by individuals, it does not exist as an independent entity which can be appealed to in an objective manner, but rather represents the adaptation to experience of individual humans. This being the case, 'reason properly used' (Hayek 1988: 8), or reason with a small r, is reason made effective by an awareness of its limitations. Rather than irrationalism, critical rationalism is 'not an abdication of reason but a rational examination of the field where reason is appropriately put in control' (Hayek 1960: 69). Critical rationalism is constitutionally aware that reason is limited in its scope, experimental in its nature and only possessed in a partial or limited sense by individuals. Moreover, critical rationalism recognizes that non-rational modes of behaviour, such as habits and skills, play a central role in the success of our actions.

Popper picks up this argument about the crucially limited nature of reason and uses it as the basis of his two theses of the core of the critical rationalist approach. He writes:

i) we are fallible, and prone to error; but we can learn from our mistakes.
ii) We cannot justify our theories, but we can rationally criticize them, and tentatively adopt those which seem best to withstand our criticism, and which have the greatest explanatory power.

(Popper 1972: 265)

Reasoning is immanent criticism: it is a process of trial and error, conjecture and refutation (Popper 1989: 51), based on the notion that we learn from our mistakes. This is reason properly understood. It is to say: 'that rationalism is an attitude of readiness to listen to critical arguments and to learn from experience' (Popper 1966 vol. 2: 225). Reason is critical argument over theories about experience and, as with Popper's conception of 'world three' and objective knowledge, so with his notion of the advance of reason; it is the current state of critical debate. What he advances is an 'interpersonal theory of reason' (Popper 1966 vol. 2: 227), that views reason as a critical debate within a tradition of thought, rather than as an isolated and abstract process of deduction. Again this stresses the cultural or social nature of reason, the process of reasoning takes place within a social context of critical debate. As with the advance of objective knowledge, that which it is 'rational' to accept is the theory that has survived criticism and testing up to the present time. In this 'natural selection of hypotheses' (Popper 1972: 261) we pursue a process of trial and error, with error elimination through the feedback mechanism of experimentation and criticism, which does not provide us with eternal truths, but rather allows us to avoid making the same errors in the future.

Throughout this focus on the key role of inter-individual debate, critical rationalists maintain a methodological individualist outlook and are clear that they do not view reason as an entity detached from the individuals who undertake it. Hayek argues that reason is not eternal and objective even in this procedural sense. Rather he argues that the human mind is a product of culture. Reason is a product of the wider process of the evolution of culture as it affects the human mind. Cultural evolution and cultural selection have shaped human rationality (LLL vol. 3: 157, 166) and as a consequence reason is a social product as well as a social process; and it ought not to be considered as apart from the social conditions which generate it (Hayek 1960: 38).[28] Social institutions ought not to be approached as though they were the product of deliberate rational design: this is an error because reason developed in tandem with these institutions in a social context.[29] We did not acquire reason and then shape other social phenomena: rather we were able to develop reason in part because we developed other social institutions that supported its exercise.[30] This means that social practices existed and functioned before humans were aware of the nature of the role which they fulfilled. It was not reason and understanding of the purpose of the practice that led to its repetition but rather its successful serving of some function. All this leads Hayek to his conception of the true role and focus of social science:

Most of these [legal] rules have never been deliberately invented but have grown through a gradual process of trial and error in which the experience of successive generations has helped to make them what they are. In most instances, therefore, nobody knows or has ever known all the reasons and considerations that have led to a rule being given a particular form. We must thus often endeavour to *discover* the functions that a rule actually serves. If we do not know the rationale of a particular rule, as is often the case, we must try to understand what its general function or purpose is to be if we are to improve upon it by deliberate legislation.

(Hayek 1960: 157)

Functional explanations are not, in Hayek's view, deterministic, they do not proceed in a manner akin to the historical laws of historicism which, once discovered, can serve as a guide to the prediction of the future. Rather social institutions are formed by individuals who are unaware of their function – and in this sense are purposeless – yet in some sense are aware of the beneficial results which arise from them. Hayek advocates an approach whereby the social scientist seeks to discover the function of social institutions in order to understand them. This approach rejects anthropomorphic interpretations of interpersonally generated institutions and, instead, seeks to understand why these institutions and practices have persisted. Like the Scots before him, Hayek operates by assuming that practices that persist fulfil some function with a degree of success. There is a 'utility'[31]-based selection of behaviour that decides between the functional efficiency of human practices. However, as Hayek notes, this sense of utility is not 'known to the acting persons, or to any one person, but [is] only a hypostatized "utility" to society as a whole' (LLL vol. 2: 22).[32]

More significantly, human practices can perform their function without knowledge of that function being possessed by those who act in accordance with the practices. With Hayek's notion of utility he is able to assert that human practices exist and function to allow the survival of the group long before that group is aware of the function that they serve (LLL vol. 1: 75). As a result, he believes, human civilization is built on the functional efficiency of practices that are followed without being understood. Hayek seeks to understand civilization as a 'functioning order' (LLL vol. 2: 98), whereby individuals adjust to complex circumstances by making use of practices that function to preserve that order. Our institutions are the adaptation to circumstances by individuals who possess only limited knowledge: they are adaptations to our ignorance.[33] Moreover, these rules and institutions embody knowledge distilled from past experience that is not apprehended in a conscious manner. For example we follow the rules of morality not because we are conscious of the beneficial results that will arise, but rather because we have some sense of the importance of obeying the rules themselves. The process of rule following is not deliberative, but is rather the result of a process of cultural evolution. As the following chapters progress we will see how Hayek advances this approach by noting how humans have formed rules in reaction to circumstances which allow the functioning of a social order. Hayek lays a great deal of stress upon what he refers to as the 'twin ideas of evolution and the spontaneous formation of order' (Hayek 1984: 177). In terms of social theory what he means by this is that society is a spontaneous order and that, as spontaneous orders form and change through an evolutionary process, so social change ought to be understood as a process of evolution.

We are able to draw a number of common themes from the conception of science as a spontaneous order concerned, in social science, with unintended consequences. First, the approach depends on a particular conception of what it is to be human: that humans are order-seeking beings whose minds are classificatory and, further, that such mental classification is originally non-deliberative but becomes the model for the conscious act of science as humans seek stability of expectations. Second, the pursuit of science is the examination of problem situations which arise from the conventionally generated 'state of the debate' equilibrium and, with relation to the social sciences, the objects debated are subjectively theorized concepts. Third, that the subject matter of social science is primarily the study of the unintended consequences of human action and that this leads to a preference for functional explanations acquired from conjectural history. Fourth, that scientific debate evolves by trial and error or through the immanent criticism of a tradition by those scientists socialized within it. Finally, that as science is a debate, a degree of freedom is justified from considerations of epistemological efficiency in order to facilitate both adjustment and discovery. The modern writers, like the Scots before them, have developed their own understanding of the practice of science and social science. They have criticized rival approaches which they term constructivist rationalism, and have advanced instead a spontaneous order approach which they refer to as critical rationalism. We may now proceed to examine how they apply this spontaneous order approach to the explanation of social institutions.

7 The evolution of morality

General rules and stability of expectations

Hayek believed that humans are essentially a 'rule-following animal' (LLL vol. 1: 11). If the human mind is a system of general rules adapted to experience and typified by the classification of phenomena then this shapes a great deal of human behaviour and leads to a propensity to develop – both deliberatively and non-deliberatively – rules of behaviour.[1] He argues that human integration in a social context is not the result of an association to serve common goals, but rather is to be understood as being the result of rule following by individuals (Hayek 1978: 85). Rules, Hayek notes, are necessary to create any type of order, in the sense that they are regularities that allow mutual adjustment and adaptation. As noted in the previous chapter, he cites the example of crystal formation whereby particle adjustment under general rules determines 'the general character of the resulting order but not all the detail of its particular manifestation' (LLL vol. 1: 40). In this sense, while the abstract entity remains the same, the crystal, the adjustment of particles in line with the general rules in particular circumstances shapes the precise form the crystal takes. In the case of spontaneous orders such rules are not commands issued with the conscious intention of creating a specific particular form (LLL vol. 2: 14), rather they are rules which facilitate the formation of the order itself. This relates to Hayek's assertion that the social order has no purpose, but rather serves the purposes of its individual constituents. The function of the general rules that facilitate the social order is to serve the purposes of the individuals concerned as they adjust to each other and their circumstances in order to form the order itself (LLL vol. 3: 109). Such general rules are formal in the sense that they are purpose independent and apply generally to all particles in a given situation.[2]

Hayek defines a rule as: 'a propensity or disposition to act or not to act in a certain manner' (LLL vol. 1: 75). Generalized rules deal with 'kinds' (LLL vol. 2: 22) of behaviour, they specify conditions for the formation of an order rather than act as a deliberate attempt to secure a particular manifestation of that order. Hayek refers to the resultant order as an 'isonomy' (Hayek 1960:

164), or rule-formed order whose particular form is the result of particle adaptation to those rules. General rules in this sense are universally applicable to all the relevant particles/parties. Indeed, a measure of the generality of a rule is precisely that it takes this universal form.

Hayek believed that humans, as rule-following animals, have built civilization upon the practice of forming and obeying general rules. It is his central assertion that we exist in a framework of rules which act to facilitate civilization and which we have made but which we do not understand (Hayek 1988: 14). Moreover, general rules do not provide us with certainty, merely with a species of probability or stability which facilitates the interaction of particular individuals while always leaving open the possibility of there being unexpected consequences of human action. Such a concern for stability can be traced, as we saw, to Hayek's view of the human mind as an order of classification of experience which humans make use of to understand their circumstances. Understanding is ordered classification and for this reason, Hayek argues, humans are most comfortable with experiences that fit their established order of classification. Humanity has developed mental orders and cultural practices that ease understanding and reduce uncertainty and fear of the unpredictable. So long as we follow these established practices in a given context the world is fairly predictable (Hayek 1967: 81). It is this function, the reduction of uncertainty, which is the key role played by much of our social behaviour.

As general rules – or social phenomena that operate through the formation of general rules – reason and habit react to the complexity of social situations and the limited knowledge of individuals, in order to facilitate the mutual adjustment of individuals to their circumstances and to each other. This being the case general rules are not only reactions to our ignorance, but they are also developed and refined, they evolve, as our experience advances (Hayek 1960: 66). When rules are formulated in a general manner they apply to an unknown number of future cases. They stabilize our expectations in these future situations by placing constraints upon the details of the circumstances such that we can adjust our behaviour with a reasonable hope of success. This is not, however, to say that they must remain eternal and changeless. General rules are subject to refinement as experience grows.

A generalized rule such as habit, custom or law, provides a stability of expectation that increases the knowledge of individuals by ruling out certain possibilities in given future situations.[3] Law and general rules provide data, they function to communicate information. In terms of the order produced by rule following both Hayek and Oakeshott are keen to stress the difference between what they refer to as 'Cosmos and Taxis', 'nomocracy and teleocracy' or 'enterprise and civil' association.[4] Their main point is that a rule-governed order, a civil association or nomocratic cosmos, possesses 'no extrinsic substantive purpose' (Oakeshott 1990: 110). It is not an association of individuals linked by a common purpose or in the pursuit of a specific end, but rather represents individuals linked together by shared regularities of behaviour (Hayek 1978: 74; Oakeshott 1990: 112). The order that arises as a result of adherence to generalized rules is an unintended consequence of the adherence to those rules, it is a spontaneous order; and while the rules function to preserve the order they are not adhered to with this end in mind.

Habit, custom and tradition

As we observed when we discussed Hayek's psychological theories, he considered the human mind to be an order that develops in reaction to experience of the surrounding circumstances or environment. The order of the mind is shaped by a process of classification of the environment through which new events are interpreted in the light of past experiences. We noted how Hayek viewed science as an attempt consciously to replicate this ordering process in a hypothetico-deductive manner. There are, however, other forms of human mental ordering which are equally as important as science in the framing of human knowledge, but which do not occur in a deliberative manner. Chief among these is the psychological phenomena of habit. Habit is a species of mental conditioning (Hayek 1976: 87), whereby generalizations and classifications are developed in the mind in a non-deliberative manner. Habitual rules of behaviour are the result of experience and develop over periods of time. They embody human knowledge in the sense that they are lessons drawn from experience, lessons, however, which are learned unconsciously. Hayek asks: 'is *knowledge* involved when a person has the habit of behaving in a manner that, without his knowing it, increases the likelihood that not only he and his family but also many others unknown to him will survive – particularly if he has preserved this habit for altogether different and indeed quite inaccurate grounds?' (Hayek 1988: 139). His answer is an unequivocal yes.

Habits embody knowledge: they provide non-deliberative rules of behaviour which act as rules of thumb or guides to humanity's relations to phenomena which resemble those of past experience. For Hayek habits are a form of general rule drawn from experience that deal not with specific factual observances, but rather with 'kinds' of ways of acting in recurrent situations. Habituation is a non-deliberative process, which is to say that habits are not deliberately acquired nor are they consciously formulated. They represent abstractions or generalizations from experience which are deployed in an unthinking, second-nature like manner when situations, or similar situations, recur. The interesting feature of habits, for Hayek, is their non-deliberative nature: they are regularities that are followed but never properly verbalized or indeed consciously adopted. He writes:

> That such abstract rules are regularly observed in action does not mean that they are known to the individual in the sense that it could communicate them. Abstraction occurs whenever an individual responds in the same manner to circumstances that have only some fea-

tures in common. Men generally act in accordance with abstract rules in this sense long before they can state them. Even when they have acquired the power of conscious abstraction, their conscious thinking and acting are probably still guided by a great many abstract rules which they obey without being able to formulate them. The fact that a rule is obeyed in action therefore does not mean that it does not still have to be discovered and formulated in words.

(Hayek 1960: 149)

Habitual thought is generalized but not in the specific theoretical manner which the conscious pursuit of science engenders. Habits represent tacit pre-suppositions, expectations in the sense that they are taken for granted, which reduce uncertainty and stabilize expectations by guiding individuals' reaction to their environment.

Following Ryle, Hayek stresses the non-deliberative nature of habitual behaviour by drawing a distinction between 'knowing how' and 'knowing that' (Hayek 1988: 78). The point of this distinction being that the success-ful functioning of a generalized rule of behaviour does not depend on any conscious apprehension of its overall function. He couples this approach with Whitehead's observation on the importance of non-deliberative behavi-our for the progress of civilization (Hayek 1960: 22; 1979: 154; 1984: 221). This argument is intended to display the vital 'supporting' role played to civilization by the non-deliberative following of habituated generalizations. The more we are able to achieve without deliberative thought, the more of our mental capacities are freed up for deliberation on other matters.[5]

Though habits are unintentionally acquired regularities of behaviour they are not eternal, nor are they immutable. Habits survive, in Hayek's view, because they succeed. They allow an efficient reaction to the environment which permits the survival of the holder of the habit. This feeds into Hayek's theory of cultural evolution, in the sense that successful habits assist the survival of both individuals and groups of individuals. Habits exist in a space 'between instinct and reason' (Hayek 1988: 10–11, 21): they are neither innate, in the sense that they are acquired, nor are they rational, in the sense that they are not deliberatively acquired. In this way habits, and their group-level equivalents customs and traditions, exist as 'tools' (Hayek 1960: 27) which individuals make use of in a non-deliberative manner. Therefore if we are to study such phenomena it must be a study that seeks to observe their functioning without any presupposition of a rational or inten-tional purpose.

Hayek also follows the Scots by arguing that humans are naturally sociable animals. For Hayek humans are sociable, but they are also socialized into a particular group. They are, as a result, socialized into the habitual or customary and traditional behaviours of that group (Hayek 1988: 12). The habitual or customary rules, which make social life possible, form the basis of human societies and are the origin point of our notions of law and government. Hayek argues that socialization, like habituation, is an innate propensity of human beings. The acquisition of rules of behaviour by imitation represents a key source of the successful transfer of experience and knowledge necessary for the survival of the group and the species. Indeed, socialized traditions form the backdrop to all human activity. Customary behaviour is learned by imitation, not conscious imitation, but rather a non-deliberative process which mirrors that of habit formation. The successful transmission of habits through socialization ensures the persistence and survival of groups of humans (Hayek 1988: 16). In this way, even reason, which Hayek noted is a cultural product, is dependent on this process. He argues: 'In a society in which rational behaviour confers an advantage on the individual, rational methods will progressively be developed and spread by imitation' (LLL vol. 3: 75). Further, there exists an implicit filtering device in this process of socialization in that, for Hayek, it is the cultural transmission of successful practices (Hayek 1984: 324). Only successful groups with successful traditions will be able to produce offspring who will survive long enough to be socialized: therefore only practices which encourage human survival will become customs which are transmitted over long periods of time. This evolutionary process lies at the heart of Hayek's descriptive theory of social change. Socialization though, as with habit, has no explicit moral criteria. People can be socialized into 'bad' practices, as the Scots noted with reference to infanticide, but they will only continue to be socialized into a practice if it fulfils a function that allows group survival. If the circumstances of the group change, then the practice, though it may proceed for several generations, will either be discarded or adapted to the new circumstances in a manner that better functions to support the survival of the group (Hayek 1978: 10).

Custom and tradition play a vital role in the successful survival of humanity. While habit operates on an individual level in reaction to the environment, custom operates through socialization on a group level. A custom is a habitual convention among a group of individuals. As Hayek puts it: 'the existence of common conventions and traditions among a group of people will enable them to work together smoothly and efficiently with much less formal organization and direct compulsion than a group without such common background' (Hayek 1984: 147). A custom is a group habit, a habitual practice or convention of behaviour among a group. While customs stabilize expectations by providing a degree of predictability to human actions they function, as with habits, in a non-deliberative manner. We act according to custom before we have any idea of why we do so; indeed a custom is rarely, if ever, deliberatively developed or applied. It operates largely with un-organized social pressure acting as the enforcement agent (LLL vol. 2: 34). People are expected to act in the expected manner, to act with propriety, and face disapproval or social exclusion if they do not.[6]

The acceptance of customary behaviour does not depend on a rational understanding of the utility of the practice, rather it is believed in not because of the results which it produces but because of its nature as a custom into which people have been socialized (Hayek 1978: 85). The 'done thing' is the done thing because it is done and not because the doer has decided that it works well. Hayek points to the concept of a taboo as an example of this. Taboos are an expression of negative customary knowledge. They restrict behaviour without the need to experience the rationale behind the restriction (Hayek 1978: 86). Customs largely exhibit negative knowledge in the sense that they restrict behaviour to a particular path without conscious awareness of the social function of the practice. We do not know the precise results of deviating from customary behaviour: all that we know is that custom and propriety dictate that we not do so.

The moral values of a social group exist and are held in a largely non-deliberative manner that is socially transmitted through the process of socialization. Morality is a customary tradition that embodies knowledge beyond that which any individual is capable consciously of formulating. Hayek notes:

> the fact that the tradition of moral rules contains adaptations to circumstances in our environment which are not accessible by individual observation or not perceptible by reason, and that our morals are therefore a human equipment that is not only a creation of reason, but even in some respects superior to it because it contains guides to human action which reason alone could never have discovered or justified.
>
> (Hayek 1984: 320)

Customary conventions of behaviour are the embodiment of generations of trial and error experience that constitute groups' notions of right and wrong. Tradition is the transfer of this knowledge gained from experience in a non-deliberative manner. The notion that a form of behaviour is not the 'done thing' arises because in the past it has been done and proved harmful.

Cultural evolution

Hayek lays a great deal of stress upon what he refers to as the 'twin ideas of evolution and the spontaneous formation of order' (Hayek 1984: 177). In terms of social theory what he means by this is that society is a spontaneous order and that, as spontaneous orders form and change through an evolutionary process, so social change ought to be understood as a process of evolution. As we noted in the introduction, Hayek argued that Darwin's theory of evolution was strongly influenced by the cultural evolutionary models of the Scots. Theories of social evolution existed long before Darwin's biological appropriation of the approach. Moreover, Hayek's theory of evolution differs in some marked respects from that of Darwin. He argues at length that though Smith influenced Darwin's approach, the model of cultural evolution to which he alludes contains no genetic or biological implications

(Hayek 1988: 23–5). It is not biological humans who evolve in this model, but rather their knowledge and culture. It is not a case of individual level survival of the fittest, but instead, of group level evolution of cultures through adaptation to circumstances and the efficient use of knowledge.

Hayek argues that the general rules that allow a social order to form are the product of cultural evolution rather than deliberate institution (Hayek 1967: 243). Such rules and institutions embody the knowledge of circumstances garnered by past generations; they are transmitted as traditions rather than as deliberate understanding of the function that they serve. The evolution of these practices displays the evolutionary 'growth' or development of human knowledge. Behind Hayek's functional analysis is a model of evolutionary epistemology (Hayek 1988: 10), a model which he compares directly to the rendering provided in Popper's theory of science (Hayek 1978: 43).[7]

Cultural evolution is about the transmission and adaptation of knowledge, beliefs and customs: it refers to the evolution of the cultural heritage of a people especially in relation to habits and customs, which, as we have seen, are characterized by the human propensity to classify experience according to rules. Hayek follows Hume in describing this process through a metaphor of path formation (Hayek 1979: 70–1).[8] He argues that humans are socialized into a particular cultural tradition which represents an adaptation to circumstance, and that the development of that tradition represents the adaptation to changes in those circumstances. Cultural evolution proceeds by an 'experimental process' (Hayek 1988: 46) of trial and error adaptation to circumstances with successful practices being repeated, and through repetition, becoming habitual. As we noted before with relation to the trial and error advance of science, such evolution cannot be planned or predicted. Cultural evolution proceeds by a process of reaction to the unintended consequences of human action. Moreover, it depends upon both adaptation and the transmission of successfully adapted practices. Which is to say that as well as being based on a growth of knowledge it is also based on the communication of such knowledge to other members of the group. Cultural evolution comprises not only the evolution of knowledge in reaction to circumstances, but also the gradual aggregation of knowledge as new practices and classifications of phenomena are discovered and absorbed. Hayek writes:

> Cultural transmission has however one great advantage over the genetic: it includes the transmission of acquired characters. The child will acquire unconsciously from the example of the parent skills which the latter may have learnt through a long process of trial and error, but which with the child become the starting point from which he can proceed to greater perfection.
>
> (Hayek 1978: 292)[9]

Cultural evolution is the 'selective evolution of rules and practices' (LLL vol. 3: 154) based on a process of 'winnowing or sifting' (LLL vol. 3: 155) grounded in the comparative success of groups which adopt differing practices.[10] There are two parts to this concept of evolution, there is the change or adaptation of a particular tradition of rules, and there is the survival of groups who hold those rules in competition with other groups. Unlike biological evolution this process can occur relatively quickly in the sense that it differentiates humans as a species to a far greater degree than biological characteristics (LLL vol. 3: 156). As we noted before, Hayek regards the mind as a cultural product so, it seems, when he talks of cultural evolution he is talking about the evolution of the human mind within the context of a particular cultural tradition. As he puts it: 'Cultural selection is not a rational process; it is not guided by but it creates reason' (LLL vol. 3: 166). It is the evolution of successfully functioning practices that are adapted to the circumstances in which the social group exists that represents the operation of a 'successful' tradition. Chief among these adaptations is the system of moral rules which, like Hume's path, embody knowledge and guide action by ruling out unprofitable behaviour. Different groups of humans develop different cultural traditions that they transmit to successive generations. However, the groups that develop the most efficiently functioning traditions hold a comparative advantage over other groups with whom they compete. Hayek illustrates this by calling on Smith's analysis of the division of labour, arguing that the successful cultural evolution of the division of labour gives the group in question a marked advantage over other groups and leads to a process of 'group selection' (Hayek 1988: 120).

Hayek argues that Smith discerned the phenomenon of group selection but failed to develop it within his work (Hayek 1967: 86, 100). He also points out that there is a strong sense in Hume's writing that practices which aided human survival, survived themselves and were communicated to succeeding generations. Practices which did not aid survival, or which faced alternative practices which were more efficient in aiding survival, were superseded. This, in Hayek's view, is what Hume meant by utility; not a deliberative calculation but rather a successful adaptation retrospectively discerned only because those who held it succeeded (Hayek 1967: 114). Group selection is based on a non-deliberative functionalism. Successful groups do not know that they prevail or why they prevail, they simply do so (LLL vol. 2: 21, 145). It is the persistence of the systems of rules that they develop, rules that constitute their identity as a group, which indicates that they operate in a successful manner.

There is an evolution of knowledge within a particular tradition and a selection of traditions between different groups. These two senses are highlighted when Hayek argues that the transmission of cultural practices occurs on an individual level while the selection of systems of practices occurs on a group level (Hayek 1967: 67). Individuals adapt to their own unique local circumstances and the development of shared, social, practices represents a group reaction to its environment. Successful practices are transmitted between members of the group and come to form the cultural tradition through which the group operates and is constituted. In Hayek's terms:

> the properties of the individuals which are significant for the existence and preservation of the group, and through this also for the existence and preservation of the individuals themselves, have been shaped by the selection of those from individuals living in groups which at each stage of the evolution of the group tended to act according to such rules as made the group more efficient.
>
> (Hayek 1967: 72)

Individuals and groups survive by adapting their behaviour to changes in circumstances and in relation to the growth of their experience of those circumstances. A successful tradition is one that has adapted practices to allow social interaction that encourages the survival of the group and its individual members.

However, there is also the process of competition between groups with different cultural traditions. Hayek argues that, as human civilization evolves, groups with poorly functioning rules fail and are absorbed into, or superseded by, other groups (Hayek 1960: 36). Groups with successfully functioning rules will prevail over other groups, not because the rules themselves have any intrinsic worth, but rather because they facilitate the maintenance of an extended order between the members of the group (Hayek 1967: 68). The extended order of particular groups, their shared practices and similar behaviour, exist because those practices have displaced practices which did not function to preserve the order: and those groups have displaced groups with these unsuccessful, 'malfunctioning' practices (Hayek 1967: 70).[11] Group selection, as an element of cultural evolution, is the selection between groups in terms of the success or otherwise of their systems of rules (LLL vol. 2: 22). The most obvious factors in group selection are those practices which succeed in allowing reproduction because they guarantee the survival and extension of the group.

If, as Hayek argues, the market and the division of labour are viewed as giving groups a comparative advantage in terms of cultural evolution, then it is as a result of the fact that trade and related cultural phenomena encourage population growth (Hayek 1988: 39). As Smith noted, the division of labour is responsible for the growth of population and is necessary for population to be maintained at current levels let alone in order for it to grow (Hayek 1991: 74).[12] Economic progress, as a result of the division of labour, the market and trade, has increased the population of those nations that develop them to a level whereby without them their inhabitants would starve and die (Hayek 1978: 19). Also, following Smith, Hayek notes that our civilization, dependent as it is on the division of labour and specialization, depends on the existence of cities (Hayek 1960: 340–1). The existence of cities, in Hayek's view, allows millions to survive who would otherwise perish. The division of labour affected in urban areas is the key to economic growth and to the consequent growth in population. Population density in cities allows the development of specialization and draws people to the urban areas in the hope of securing employment in industry (Hayek 1988: 40). As specialization leads to a growth in specialist knowledge so the growth of population acts to prompt the process further by increasing the number of possible specialists. It is this growing diversity of specialist knowledge rather than the increase in individual intelligence which supports economic progress.

Group selection between cultural traditions is a result of the number kept alive, and socialized within, the tradition. As Hayek puts it: 'For the numbers kept alive by differing systems of rules decide which system will dominate' (Hayek 1988: 130). The only groups that survive are those which possess customs which function to provide for reproduction and the raising of children (Hayek 1988: 84).[13] Civilization, and the survival of the group, depends on the survival of children and upon the use of knowledge and cultural practices that allow this. Put another way: 'The size of the stock of capital of a people, together with its accumulated traditions and practices for extracting and communicating information, determine whether that people can maintain large numbers' (Hayek 1988: 124). This is, in Hayek's view, essentially the same argument as that advanced by Hume and Smith with regard to the development of property rules (Hayek 1984: 321–2).[14] Moreover, as the Scots also note, successful cultural practices are often spread by immigration and colonization. Hayek notes that population growth is often as much a product of immigration and the absorption of less successful groups, as it is a result of the increasing size of families (LLL vol. 3: 159).

Population may, however, be the closest point of affinity between Hayek's cultural evolution and Darwin's biological evolution. Hayek argues that, in a very real sense, humanity's purpose is survival and reproduction (Hayek 1988: 133). This, however, is not a moralized argument. Population growth is not good in any moral or ethical sense, but rather is a descriptive measure of the successful functioning of practices that secure a human 'biogenic' drive. The selection of groups through a standard of success based on population is not a moral argument for the 'goodness' of the group's practices: it is a practical measure of their success that aims to explain the evolution of cultural traditions. As Hayek notes, we have become civilized in order to rear more children and not because of any intrinsic moral value in the individual or the practices which they make use of (LLL vol. 3: 167–8). Such a functional understanding of cultural evolution makes use of a sort of 'calculus of lives' (Hayek 1988: 132) where success is measured in terms of population growth.

It should be noted, however, that the idea of group selection is not uncontroversial. Dawkins and others have attacked group selection in biology as flawed because it fails to provide an explanation of the link between the individual and the group. Denis (1999: 3) and Vanberg (1986:

83–6) take up this view in relation to Hayekian group selection. They argue that 'free rider' problems prevent group selection theories from convincingly explaining how group level advantage can differ from an aggregation of individual level advantages. Rules that encourage group survival are not necessarily the same as those which benefit individuals.[15] Group selection theories have a problem explaining why individuals would submit to rules which are not to their immediate personal advantage. The obvious answer to this, as exemplified by Ferguson's argument over the willingness of individuals to die to protect the group, is that individuals are socialized within groups. In order for there to be free riders there must be something for them to 'ride' upon: a social group constituted by a set of commonly accepted conventional rules. This being the case, if such groups are constituted by a rule that restricts free riding, then the critique fails. Sociability itself, as well as being the basis for the generation of cultural rules, is the glue that allows group advantages to exist.

This leads to a second problem with Hayek's group selection. As we have seen, Hayek uses population as an indication of group success, but it is not immediately clear why individuals should care about population growth.[16] The answer to this is that they don't. Population is purely indicative of functionality and is in no way a conscious rationale for the selection of rules. To the extent that individuals are aware of the population as significant it is on the micro-level of wishing to keep those close to them alive.[17] A more significant criticism of group selection lies in the accusation that, as with all functionalist explanations, it is holistic in character and undercuts a commitment to methodological individualism in social science.[18] If Hayek wishes to advance a group selection argument, then he must be able to provide a link between group and individual levels of selection that keeps his methodological individualism intact. We have already considered the free rider problem by referring to sociability and the fact that rules are not chosen on the basis of 'rational self-interest', what we now require is a conceptual link between individual level evolution and group level selection. This problem appears to be compounded when we consider that it is Hayek's view that it is systems of rules and not individual rules or individual applications of rules that are significant for group selection. As Vaughn has noted, even if group selection is accurate it occurs on a systemic level, which is to say that 'bundles' of rules survive (Vaughn 1984: 124). Such customary 'bundles' are indeed group level phenomena but, as we have argued above, they evolve through a medium of individual selection. However, as Hodgson notes: 'While group selection is occurring, *individual* selection is also going on simultaneously within the group' (Hodgson 1991: 74, his italics). Individual experiments in living within the broader group tradition are the key to successful adaptation – with the link preserving the group being provided by socialization and imitation.[19] This argument provides an instrumental justification of freedom similar to the one we observed in Chapter 6 on science whereby it is understood as freedom for adaptation in order to improve the efficiency of the tradition in a piecemeal manner. Freedom to adjust exists within the broader confines of the tradition of moral behaviour. Individuals imitate those of their fellows whom they consider to be successful within the context of the group. Moreover, groups import practices through the emulation of other groups. This group selection can be understood not by the 'death' of group members but by the disappearance of the practices which constituted the identity of the group. The notion of the total disappearance of a social group, whether by death or total absorption into a more efficient group, is a concept that refers to the earlier stages of cultural evolution (Vaughn 1984: 125). Communication between groups and emulation of technology and knowledge means that relatively unsuccessful groups are able to maintain group coherence as evolution progresses (Witt 1994: 184): they are able to imitate and to adapt the practices of more successful groups. It is for this reason that the population indicator becomes less significant in more economically developed cultures. The possession and accumulation of human capital, in the sense of knowledge, becomes a more significant indicator of group success than sheer population size (Radnitzky 1987: 24). Minogue (1987), then, is correct to note the shift between the earlier and later stages of cultural evolution; group selection indicated by population levels typifies functional efficiency during the earlier part of the process while experiments in living and imitative reform suggested by the success of other groups typify functional efficiency in the more advanced phase of cultural evolution.

Leading on from his focus on population Hayek engages with a problem which, as we have seen, much exercised the Scots. He cites Carr-Saunders with approval (LLL vol. 1: 148–9), and notes that there is a sense in which our moral rules are selected in relation to their conduciveness to survival and to the sustainable growth of population (LLL vol. 3: 160–1). As the economy grows and increasing numbers are supported, humanity will develop or adapt those practices aimed at population limitation. A change in circumstances, in this case economic growth, will render practices such as infanticide obsolete by reducing their economic 'logic'. This, however, is not a deliberative process and it rests on two principles: first, a practice such as infanticide may pass out of use because it no longer fulfils its social function (for example, economic growth renders it unnecessary); and second, the growth of knowledge (of contraceptive practices) results in the function of infanticide being more efficiently fulfilled by other practices developed from experience.

To say all this, however, is not to argue that practices such as infanticide are placed beyond criticism. Rules are indeed selected on the basis of the survival of the population, but to explain the function of a practice in relation to this need not imply approval. Like the Scots' writings on the ancient Greeks, we might understand the function yet disapprove of the practice adopted and criticize its continuance after the conditions that entailed it have passed. Our moral practices are adaptations to our circumstances, and changes in circumstances may render them obsolete and make them abhorrent in the eyes of future generations. However, as Hayek (1988: 152) notes, practices such as the Eskimo exposure of the elderly were developed to ensure group survival, and unless the group survived by their successful functioning, there would be no succeeding generations to disapprove of the practice.[20] Hayek also notes the significance of emotion in this process. He notes the historical example of Roman magistrates who were praised for condemning their own children to death and argues that 'we have learned to avoid the gravest of such conflicts, and in general to reduce the requirements of formal justice to what is compatible with our emotions' (LLL vol. 2: 148).

Should circumstances dictate a practice which runs against the grain of the human emotional attachment to offspring, then such practices will tend to be replaced by more emotionally acceptable approaches as circumstances permit and as knowledge grows. We adapt to circumstances and change our moral practices in line with changes in those circumstances: though, as we noted, the habitual and customary nature of moral practices means that this change is subject to 'evolutionary lag' (Gray 1986: 50) and is gradual and piecemeal. A custom is only an obstacle when it is no longer the only, or the most efficient, way of doing something, or fulfilling some function. It is for this reason that Hayek attacks conservatism as a failure to recognize change as a positive force that increases the successful functioning of the group in reaction to changes in circumstances (Hayek 1960: 400). As we noted before, habit and custom are flexible and allow gradual changes in line with changes in circumstance. The 'licence to experiment' is not based on whole-sale reform of the social system. Instead change occurs when tolerance is granted to 'the breaker of accepted rules' who has demonstrated willingness to adhere to 'most rules' in exchange for the opportunity to act as a 'pioneer' in a particular field (LLL vol. 3: 204).

Our habits are ingrained but they admit change and they must adapt to changes in circumstances in order to facilitate the survival of children. To this end Hayek adopts an argument similar to that of Mill's 'experiments in living' (Hayek 1960: 127). Experiments in living and the success or failure of such provide the examples that guide the changes in human habits and cultural practices (Hayek 1960: 36). If evolution depends on adaptation to circumstances, then freedom is a key factor in the ability of individuals to adapt successfully. The argument in favour of freedom is precisely that it enhances the efficiency of adaptation.[21] Thus Hayek describes how evolution occurs and then argues that given the reality of this model of social change it is preferable to encourage individual freedom under general rules to ensure efficient adaptation. Reform of our moral traditions must proceed by a process of immanent criticism (Hayek 1988: 69). We draw on our reason to examine the cultural practices of our group and assess their functionality. One criterion for selection is the compatibility of a rule of behaviour with the other principles of the cultural tradition; we seek to weed out contradictory practices in order to stabilize expectations and reduce confusion. Infanticide is contrary to the principles respecting individuality, the desire for procreation and so on, therefore it can be criticized and, as circumstances allow, adapted or discarded. Such a process demands gradual, incremental and careful reform, and criticism of both the articulated and non-articulated practices which function to preserve the extended order of civilization. Those rules survive the non-deliberative process of cultural evolution that function to preserve the order, those that do not fall into disuse. Our reform and criticism of our moral tradition must be based on an awareness of this constantly changing and adapting process.

Knowledge and morality

As we noted previously Hayek's methodological individualist approach does not presuppose an assertion that individuals are by nature selfish. He rejects the notion of economic man as a universal model on the same grounds as he rejects models of perfect competition: it simply does not reflect reality. Economic man is a rationalist abstraction which, Hayek argues, is not a product of the epistemological evolutionary tradition with which he identifies himself (Hayek 1960: 61). In support of this view Hayek points out that Smith does not operate with such a model of selfish egotism – for though he argues that we seek to act in our interest, he does not assert that those interests are necessarily selfish (Hayek 1978: 268). The source of the error that attributes economic man to the Scots is, in Hayek's view, the unfortunate stress laid in the writings on the division of labour on selfishness (LLL vol. 1: 110). Hayek argues that this use of the term selfishness by the Scots is misleading: what they actually argue for is not selfishness so much as the pursuit of our own purposes, and these purposes may be either selfish or altruistic. Hayek argues that this represents a confusion of self-interest with selfishness. He argues that freedom to pursue individual aims is of equal importance to altruists and egotists (Hayek 1960: 78).

Hayek recasts the problem in epistemological terms: In order to achieve our altruistic goals ought we to consider the effects of our actions on all individuals?[22] This, he believes – owing to our limited knowledge and the unintended nature of the consequences of much of our action – is impossible. In response he advances the view that the focusing of individual attention on that individual's own interests (whether selfish or altruistic) produces the most efficient use of resources: and the most efficient use of resources to achieve human ends is in the interests of all. He notes:

> To enable the individual to use his knowledge and abilities in the pursuit of his self-chosen aims was regarded both as the greatest benefit government could secure to all, as well as the best way of inducing these individuals to make the greatest contribution to the welfare of others.
>
> (Hayek 1978: 133)

Hayek argues, persuasively if we refer back to the sections on the Scots, that the eighteenth-century conception of self-love or self-interest referred not just to individual selfishness, but rather to a concern for the self and those intimately related to one. Such self-love included love of family and friends. True individualism of this sort includes the circles of intimates around each individual. We are socialized within such circles of intimates, into families and communities, and to abstract human motivations from them necessarily renders the motivational model unrealistic. Hayek develops this view in anthropological terms to show that it served an instrumental function in the small group societies in which humans existed for millennia (Hayek 1988: 11). In such face-to-face societies the care for intimates, because of the small size of the groups, meant that the identification and pursuit of common ends was eminently possible. However, Hayek argues that we cannot transfer such models of small group emotional collaboration to the wider extended order that we have developed. It is simply not possible to extend the same emotional concern to a body of people outside our ken. Small group solidarity is an insufficient organizational principle for an extended order (LLL vol. 3: 162), but this is not to say that such groups cannot operate within that order.

For Hayek the emotional ties (love) that bind us to those close to us are concrete; they apply to particular individuals and cannot be extended to the whole of a great society.[23] We may very well love mankind, but such a love is far weaker than our care for those concrete individuals to whom we are close. Such a feature of human psychology is akin to the limitations on human knowledge that we have already seen. We feel concern for those close to us and can act efficiently to assist them through our local knowledge of their particular circumstances. However, epistemological and emotional restrictions prevent us from doing so for the care of the millions of others to whom we are related in the extended order. And the consequence of this for Hayek is that each individual should be:

> free to make full use of *his* knowledge and skill, that he must be allowed to be guided by his concern for the particular things of which *he* knows and for which *he* cares, if he is to make as great a contribution to the common purposes of society as he is capable of making.
>
> (Hayek 1984: 140, his italics)

Small group associations for common purposes exist within the extended order but do not characterize the order as a whole. Proximity and common concerns can be pursued through them in an efficient manner so long as the model is not extended to the whole of society. The great society, then, includes networks of non-economic associations.

Hayek's arguments on the epistemological and emotional constraints placed on individuals by their nature lead him to assert that in an extended order individuals should focus their attention on what they know and on that sphere which they can effectively control or influence. He argues that this is the function of the notion of responsibility. To hold that an individual is responsible for a certain sphere of action is to direct their attention towards its efficient use (Hayek 1960: 71). Prudence becomes admired as a virtue precisely because it has allowed survival and the efficient exploitation of local knowledge of circumstances (LLL vol. 3: 165). In terms of market relations in an extended order this means that we depend not on distant individuals' opinions or feelings for us, but rather on our ability to provide them with services which result from our exploitation of our own situation. This argument is related to the earlier arguments we noted in Hayek's theory of the mind and perception. He argues that we understand the actions of others by analogy from our understanding of our own actions. When we seek to understand the motivation behind the actions of others we rely upon our knowledge of our own motivations (Hayek 1976: 133). We imagine what we would do, how we would react, in similar circumstances. In this sense we attribute purpose to the action of others with reference to our own understanding of how we would act, and upon what knowledge, classification and rules we would regard as relevant in like circumstances. Our understanding of others is a product of a sympathetic (in Smith's sense) process (Hayek 1967: 58). This is a non-deliberative process on most occasions. We often cannot explain the rationale of the moral judgements that are produced by it (Hayek 1991: 151). Such moral judgements are not necessarily rationally or consciously calculated, but rather are based on a 'feeling' of what ought to be done in given circumstances. For this reason our moral code has developed in such a way that it is compatible with human feeling, with emotion.

Morality is not designed or chosen, but is based on emotional approbation. Hayek writes: 'Ethics is not a matter of choice. We have not designed it and cannot design it. And perhaps all that is innate is the fear of the frown and other signs of disapproval of our fellows' (LLL vol. 3: 167). Such a moral code, resting on approval and disapproval and the attribution of similar thought processes, leads us to assume that agents possess responsibility for their actions. As a consequence our assessment of the merit of an action is subjective, it is based on our own understanding of how we would have acted: upon our judgement of 'situational propriety' (Butler 1983: 21). Hayek again: 'The merit of an action is in its nature something subjective and rests in a large measure on circumstances which only the acting person can know and the importance of which different people will assess very differently' (Hayek 1967: 258). Moral approbation and disapprobation are the product of the fact that we cannot step into the minds of others. We assess their actions and motivations in terms of our own understanding. Thus by holding an individual responsible for their actions and judging them in terms of approval or disapproval we seek to influence their behaviour. As Hayek notes: 'We assign responsibility to a man, not in order to say that as he was he might have acted differently, but in order to make him different' (Hayek 1960: 75). Individuals exist in a social environment and they react to the reactions of others to their actions. It is because of this that we adjust our actions to that which we believe will secure the approval of others. Such a desire for approval is linked to the process of socialization and imitation, we seek approval by acting in the expected manner in given situations and, as a result, a convention is developed which represents the 'done thing' in that given situation. This inter-subjectively generated standard of behaviour represents the conventional means by which we assess the morality of behaviour. Hayek argues that a successful free society depends upon praise and blame in order to educate its members in the moral rules under which it operates: socialization requires both imitation and praise/blame to encourage such imitation (Hayek 1967: 233). Moral rules do not have their origin in a rational calculation that is subsequently consciously imposed upon a society. It is the esteem of others that acts as the inducement to follow moral rules: especially the esteem of those close to us, whom we seek to imitate and for whom we care. These conventionally developed standards of behaviour are, for Hayek, the same as the notion of propriety to be found in the work of the Scots (LLL vol. 3: 203). Such unarticulated habitual conventions are often highly difficult to express in a precise form. They are more often than not 'felt' rather than consciously deduced: like Smith's sympathy we know how we ought to behave and how we expect others to behave, but we undertake this process on a non-deliberative level. We 'feel' the right thing to do more often than we know why it is the right thing to do.

Those who deviate from the standard are disapproved of or excluded from the group and the standards become what Hayek calls, following Campbell, 'social-evolutionary inhibitory systems' (LLL vol. 3: 175). The function of the human propensity to praise and blame, to pass moral judgement, is often to affect a change in the behaviour of the subject judged and to preserve the conventional mode of behaviour that is part of the order-inducing practices of the society.[24]

Ulrich Witt has described Hayek's theory of cultural evolution as 'sketchy and unfinished' (Witt 1994: 187) and while it is true that the theory is not as developed as we might wish it to be, it appears that Hayek has provided us with the outline of a conjectural history of the origins of morality that approaches the generation of norms of behaviour in functional terms. Humans are sociable and habitual creatures in possession of limited knowledge and in search of stability of expectations. They form conventions with others which, through imitation and socialization, become a tradition of moral behaviour. This tradition adapts to changes in circumstances and evolves by trial and error on both deliberative and non-deliberative levels: which is to say that both individual experimentation and group selection play a role. Moral rules provide a species of stability of expectations by expressing the 'done thing' that is expected of group members. Moreover, these rules embody knowledge in an unarticulated form such that they direct individuals away from socially disruptive or individually harmful behaviour. Moral practice has no other function than the facilitation of mutual adjustment within the group: it is part of the invisible hand that facilitates socially benign spontaneous orders.

8 The evolution of law and government

Law

Having examined the conjectural history of the emergence of common moral beliefs we are now able to move on to Hayek's application of the spontaneous order approach to other social institutions. One of the most significant of these is the concept of private or, Hayek's preferred term, 'several' (Hayek 1960: 450) property. Several property has the function of preventing coercion and disagreement within groups (Hayek 1960: 140). By creating an individual protected sphere, defined by general rules of possession, groups prevent conflict between members over resources. Moreover, Hayek argues, such several property allows a decentralization of effort which encourages the development and utilization of local knowledge (Hayek 1988: 86). Lest we consider this the result of a process of absolutely selfish acquisition, as the development of private property is often described, Hayek stresses that though this may in part be an accurate description of the motivations of the actors, the function fulfilled is quite different. The function of several property is to secure peace within groups and to encourage the exploitation of dispersed knowledge. General rules such as those governing property provide peace and a degree of stability of expectations (LLL vol. 2: 109). Expectations are stabilized if people are provided with a clearly delineated 'known' sphere of action, thus reducing conflict and allowing peaceful exploitation of resources. General rules or conventions regarding property and contract are adopted because they fulfil these functions. Even though the persistence of such conventions does not require the conscious realization of this by those who submit to them. Several property is a product, in Hayek's view, of the process of cultural selection that we examined in the previous chapter. Its peace maintaining and economic efficiency promoting functions mean that groups which adopt it survive and flourish more efficiently than those which do not.

Thus far we have dealt with conventions that arise without the need for an institutional framework to ensure their effective operation, but as disputes inevitably arise over property claims social groups develop mechanisms to deal with them and maintain peace in the group. Following the

Scots, Hayek rejects great legislator and social contract approaches to the analysis of the origins of government on the grounds that they are historically inaccurate constructivist rationalist errors (LLL vol. 1: 10–11). Like Millar and Hume, Hayek believes that the project of instituting a system of government was beyond the scope of primitive human groups (LLL vol. 1: 97); instead he argues that social institutions such as government are not designed to serve a specific purpose, but rather evolve as spontaneous orders and fulfil a function. The institution of government is a spontaneous formation that gradually evolves from practice. Such institutions are adaptations to the circumstances and limited knowledge of primitive groups that seek to provide conflict resolution. Law for Hayek predates the conscious act of lawmaking and government is an institution which is developed to enhance and to enforce law that already exists in a customary sense. In an approach which closely mirrors that of Hume, Hayek argues that once people come to settle in one place and have selected a chief on criteria of ability (LLL vol. 2: 41), such a chief becomes the first lawmaker (Hayek 1960: 151).[1] He develops the role of 'judge-king' (Hayek 1984: 358), whose role it is to interpret existing practice to resolve problem situations. His authority to fulfil this task is based solely on the 'opinion' of those who are subject to his authority that he is entitled to wield it.[2] The role of the chief is to rule on disputes, to plug 'gaps' or to clarify customary practice in order to resolve intra-group conflicts. The chief has two primary roles which are at the heart of all governments and which have subsequently developed. The first role is to maintain order by enforcing the traditional or customary rules of the group, and the second is to issue commands to secure specific goals (such as the command of an army in war) designed to achieve communal purposes (LLL vol. 1: 76–7). These two functions of the chief are fundamentally the same as the functions of modern governments. However, the first function, of enforcing and clarifying rules to maintain order, is clearly the more important in Hayek's view, for it is this function which makes an extended society possible by resolving intra-group disputes.

The articulation of conventional practice that occurs when a chief arbitrates in a dispute is, for Hayek, the origin of law in the modern sense. He defines the nature of law as 'purpose-independent rules' that regulate the interaction of individuals and which apply to 'an unknown number of future instances' in such a way that by securing each individual a 'protected domain' they 'enable an order of actions to form itself wherein the individuals can make feasible plans' (LLL vol. 1: 85–6). Thus when he writes of a society that functions under the rule of law he is referring to a 'meta-legal doctrine' (Hayek 1960: 206) which prescribes that legislation should be conducted in line with the above definition of what makes a law. Hayek seeks to underline this point because there has been a historic confusion about the nature of law. This confusion arises from the two distinct roles that formed around the position of early chiefs. There are two 'types' of law which Hayek describes as: 'The use of enforceable generic rules in order to induce the formation of a self-maintaining order and the direction of an organization by command towards particular purposes' (LLL vol. 2: 55). The two senses of law represent two different functions of government: the formation and articulation of general rules is law properly understood, while the issuing of administrative commands is another species of legislation.

Law and legislation represent two different functions of the institution of government. Legislation develops from the execution of the commands of the legislator in the pursuit of the service functions of government.[3] Law on the other hand is generalized rules of just conduct which apply to the whole of society and whose aim is to stabilize expectations rather than to secure a specific outcome. Hayek is quite clear that law is necessary for the formation of other spontaneous orders in that it provides a stability of expectation and a regularity of behaviour which allows mutual adaptation (LLL vol. 1: 112). Laws, as Hume argued, must be general in form, known and certain, so that they can facilitate the formation of a spontaneous order by enhancing the stability of expectations. By this understanding laws are instrumental. He notes: 'When we call them "instrumental", we mean that in obeying them the individual still pursues his own and not the lawgiver's ends. Indeed, specific ends of action, being always particulars, should not enter into general rules.' (Hayek 1960: 152). Such generalized conditions applied to all provide a solid order within which individuals are able to plan their actions. Moreover, such generalized rules in the form of law form a part of the invisible hand because they serve an epistemological function. They add to our knowledge not only by stabilizing our expectations but also because they embody the experience of past generations' attempts to facilitate such an order (Hayek 1960: 157).

Hayek views law as an institution that is 'discovered' (Hayek 1993 vol. 1: 78) from prior practice rather than consciously created. In the development of the institution of government we saw how chiefs appealed to commonly held ideas about just behaviour in order to settle disputes and it is in this sense that law pre-exists its enforcement or its conscious articulation. One implication of law pre-existing its enforcement is that chiefs (and governments more generally) will only enforce, or be able to enforce, laws that are widely accepted as conforming to the extant practices of the society (LLL vol. 2: 51). Such a limitation on lawmakers implies that those laws which are articulated will exist as customarily accepted practices which have grown up out of the experience of circumstances of past generations. Chief among the experiences that will have emerged will be those which facilitate order among the social group (LLL vol. 1: 123): for example property conventions which prevent disputes over ownership. Law is 'grown' (LLL vol. 1: 95) from experience and articulated from a pre-existing convention rather than consciously created.

The articulation or discovery of law was initially undertaken by chiefs who assumed the role of judge.[4] Codification by these figures, judge-made law, does not represent the creation of law, but rather the articulation of customary conventions (Hayek 1960: 148) and the role of the judge is to provide a generalized articulation of current practice. These generalized articulations clarify conventional practice and further stabilize expectations by setting precedents. In this sense it is regularities of behaviour that are necessary to facilitate the formation of spontaneous order: the more stable the regularities, the more extensive the order that may form. The first law-making was not a conscious attempt to secure a particular set of material results, but rather was an unintended consequence of the attempt to enhance the ordering devices of the society in reaction to a particular case.[5] As Hayek puts it:

> The efforts of the judge are thus part of that process of adaptation of society to circumstances by which the spontaneous order grows. He assists in the process of selection by upholding those rules which, like those which have worked well in the past, make it more likely that expectations will match and not conflict.
>
> (LLL vol. 1: 119)

The interpretation of conventional behaviour applied to specific situations that is undertaken by judges represents the evolution of the law. Law is not designed, but rather 'It is the outcome of a process of evolution in the course of which spontaneous growth of customs and deliberate improvements of the particulars of an existing system have constantly interacted' (LLL vol. 1: 100).

Further, if law in the form of general rules is an adaptation to the limited nature of individual knowledge then this implies that lawmakers are similarly constrained by their limited knowledge. Lawmaking properly understood, in Hayek's view, always proceeds in a general manner, which is to say that its enactments are always in the form of general rules. Lawmaking, it follows, is the process of making or articulating general rules which will apply to the whole of society. And while these general rules have to be enforced in order for society to operate, the government should act to enforce them only in line with other generalized rules. That is to say, the legislative and service functions (command functions) of government ought to be circumscribed by generic rules. For Hayek the legitimacy of a government rests not only on it following its own general rules, in the form of a constitution, but in its issuing enactments which take the form of general rules (LLL vol. 1: 92–3).

Hayek gives the name 'rules of just conduct' to his conception of the procedural general rules that make social interaction possible. He defines these as: 'those end-independent rules which serve the formation of a spontaneous order, in contrast to the end-dependent rules of organization' (LLL vol. 2: 31). We should pause to note here that such a definition does not limit the rules of just conduct to law but instead encompasses the whole range of conventions regarding proper conduct. It is in this sense that the rules of just conduct are 'constitutive' of a social group, they characterize the association and are the common bond which holds it together. The rules of just conduct apply to the lawmaking function of government rather than to the service function. They have no particular end in view and act as procedural guidelines for behaviour. Rules of just conduct are, in form, abstract, general and universal. They conform to Hume's conception of general rules in that they refer to 'types' of behaviour in given circumstances rather than to the specific actions of identifiable actors. They do not command us how to act by laying down goals, and in this sense they are negative and end-independent: they are aimed at limiting the range of possible actions rather than guaranteeing a particular action. Rules of just conduct are limitations of uncertainty rather than guarantors of certainty: they stabilize expectations by reducing the field of possibilities.

Hayek believes that, like other social institutions, the rules of just conduct have been evolved in a gradual manner to help us deal with our ignorance of the complexity of our society (Hayek 1960: 66). And while such rules provide a better chance of successful action without guaranteeing results, they have nonetheless contributed to the growing complexity of society. For example, Hayek notes how rules of just conduct support trade and the market, but leave them free mutually to adjust rather than prescribe a notion of value. Rules act to reduce uncertainty, but they do not remove it entirely because they leave a degree of flexibility for individual action under the rule (LLL vol. 2: 123).

Rules of just conduct improve the chances of all by providing a guide to be considered in the formation of individual plans of action. As they do not guarantee particular results they cannot be rules of distribution or organization. They refer to commutative rather than distributive justice in that they refer to procedural regularities and not absolute certainties. Hayek argues that only situations that are the creation of human will can be meaningfully understood in terms of justice. As a result spontaneous orders that arise as the unintended consequence of human interaction cannot be considered in terms of justice (LLL vol. 2: 33). In this sense the concept of social justice cannot be justice in the same sense as understood by following rules of just conduct. The pursuit of particular, certain, distributive results cannot depend on general rules which apply to all actors because it implies treating actors in a different manner depending on a criterion of desert (LLL vol. 2: 82–3, 135). This, of course, also means, significantly, that by treating individuals in different manners in the pursuit of distributive justice we lose the expectation stabilization achieved by following generalized rules.

Hayek believes that the institution of justice evolves and is gradually discovered from the conventions of human behaviour, and that this is what is at the heart of the much-abused term 'natural law' (Hayek 1967: 101). Justice, he argues, is possible in an extended society precisely because it relies on agreement only over general rules of behaviour and not over the pursuit of specific individual purposes. We agree over rules of behaviour and need not come to an agreement over the ends of our actions within the framework these rules create. These general rules proscribe types of action but make no reference to the ends at which legitimate actions aim. Hayek terms the order that spontaneously arises from such a process as a 'cosmos' or 'nomocracy': an order which results from the adherence to abstract general rules with no agreement as to ends (Hayek 1978: 76). Thus the interplay of general rules with circumstances produces an abstract or extended order, which is just in that it was procedurally legitimate, but which has no conception of justice in the sense of desert or desired end-state.[6]

Moreover, because justice refers to ways of acting, it deals with choices about how to act, and such choices posit a notion of a responsible agent. In this way Hayek rejects end-state approaches to distributive justice because they approach the results of human interaction as though they were the results of intended action. This anthropomorphic misapprehension, which we observed earlier, neglects the decisive fact that where there is no responsible agent acting with intent there can be no meaningful discussion of justice. It is not the end-state that is properly understood as just, but rather the procedure of getting to the end-state. If we adopt the end-state view it would be possible, in Hayek's opinion, to regard the results of the market as unjust: but in order to speak of them as just we would have to attribute agency and responsibility, to view them as having been deliberately intended. The market order (on a systemic level) cannot be viewed like this, as it is precisely Hayek's point that it operates inter-individually and produces unintended consequences. It is nonsense to talk about the unintended consequences of human interaction in terms of justice.

The spontaneous social order is better understood as a framework for the attainment of individual goals. It allows the efficient use and co-ordination of dispersed knowledge to the benefit of the whole of society (Hayek 1960: 223). For example, the rules of property, as part of the framework, delineate secure individual spheres of influence that allow the efficient use of individual knowledge of particular local circumstances.[7] The framework has no active purpose of its own but rather serves the purposes of individuals. As we have seen, the framework of general rules (both non-deliberative custom and conventions and deliberatively articulated laws) facilitates the formation of order by stabilizing expectations. This process is perverted by attempts to secure particular results for particular individuals through notions such as social justice. It is not the effect in particular cases, which gives the framework of general rules its legitimacy, but the improved chances of all as a result of the universal application of the rules.[8] Such general rules are not aimed at the securing of particular human needs, but rather at the preservation of the overall order which allows individuals to pursue their needs. Justice is concerned with results only so far as they pertain to the legitimacy of the process or means of attaining them. A judge must confine his attention to the assessment of actions in terms of conformity with established rules rather than with any concern for the particular results which individuals produce under those rules. In this sense results which appear unjust might actually have arisen by a just process and so must be considered just if the general rules are to fulfil their function. The desire to correct the material situation of individuals whose lot is the result of a process that is just destroys the fabric of the system and harms all individuals by introducing a species of arbitrary action which destabilizes expectations.

In this way the rule of law, as the conditions of justice, operates in the same manner as the rules of a 'game' (Hayek 1991: 62).[9] The rules of justice specify generic conditions whose effects upon particular individuals are unknown. They become a part of the circumstances to which individuals adapt themselves, but in such a way that they are 'blind' (Hayek 1991: 76) to the particular effects of each individual's attempts at adaptation.[10] They are instrumental and like the rules of a game they introduce a degree of stability into the conduct of the players. Moreover, the rules themselves constitute the game: cricket, like society, is constituted by its rules.[11]

Similarly, as part of this overall framework which forms the rule of law, there exists a conception of a 'constitution' (Hayek 1960: 219): a conception of a law limiting the activities of government in terms of rules that function to stabilize expectations by limiting changes to those rules. This constitution lays down conditions and forms of behaviour within which the actions of government must remain. Judges are able to decide in line with the general rules of such a framework on the permissibility of the actions of a government. By limiting the scope for action held by a government in this manner the expectations of individuals are further stabilized by the prevention of arbitrary changes to the legal framework. There is, however, no need for this framework to represent a monolithic and eternal form. Though the imperfect nature of our knowledge prevents us from understanding the totality of the order which the constitutional framework supports, we are nonetheless able to reform the general character of the constitution so long as we do not seek to make it serve the purposes of particular individuals (LLL vol. 1: 41). Though many of the general rules that serve to aid the function of society may be considered to have developed as a spontaneous order from a process of unintended consequences, it is possible to attempt the gradual reformation of these rules in line with experience. This is permissible, in Hayek's view, so long as these reforms stick to the generalization criteria and continue the function of the law in providing the general conditions for the formation of spontaneous orders within society. For Hayek, progress in relation to law and government is the gradual refinement through immanent criticism of the framework of institutions that enhances the spontaneous order of society. It is possible to design conscious additions to the framework based upon experience, and it is planning in this sense, the planning of the framework, which Hayek believes is not only possible but desirable (Hayek 1980: 135). The framework of general rules that typifies the rule of law should be continually adjusted to new circumstances as they emerge and are experienced. The adjustment will not reduce stability of expectations so long as the resulting reformed rules are not retroactive and are laid down in a generalized form. All the same, such reforms must bear in mind that they are reforms to the framework, they ought to be consistent with the rest of the framework in order to facilitate the stability of expectations necessary for mutual adjustment and the utilization of dispersed knowledge.

The role of government

Having passed through a conjectural history of the evolution of social institutions we may now move on to some of the practical conclusions that Hayek draws from his application of the spontaneous order approach. He is clear that there are definite lessons to be learned from the conjectural analysis of cultural evolution.[12] The most significant lesson for Hayek is that freedom is justified as an instrument to secure certain other values that are taken for granted as being desirable.[13] The role of government in this process is viewed from this perspective through the medium of the spontaneous order approach. As we noted above, Hayek argues that there are two distinct functions that are performed by all governments. We identified the enactments which express these functions separately as law and legislation.[14] Law is the articulation of generalized rules of just conduct and legislation is an administrative command to procure a service function. While these functions are both performed by the same institution, government, he cautions against conflating them. The two functions are separate, they aim at different ends and they are enacted in different manners. This difference in function leads Hayek to suggest a constitutional model that assigns the two functions to different houses of a bi-cameral assembly (LLL vol. 3: 8). His reasoning in support of this is that the aims and enactments required to pursue each function differ, and that there is a real danger should we confuse the articulation of general rules of just conduct with the issuing of administrative commands. Hayek explains this difference by noting that rules of organization intended to secure public services are governed by a concern with securing particular results in an efficient manner (LLL vol. 3: 48). They are not concerned with the generalized formulation of rules of conduct governing the interaction of individuals. By allocating different functions to different houses of Parliament Hayek hoped to prevent a confusion of the methods required to pursue the different functions of government, to prevent the confusion of law properly understood with government edicts designed to secure a particular material result.

The more important of the two functions of government is the enhancement of the framework of general rules which facilitates the spontaneous order of society. Hayek believes that the service functions of government are subsidiary to the task of articulating the general rules of behaviour that allow society to exist. As we have noted in some detail the formation of spontaneous orders within society requires stable general rules that allow the degree of stability of expectation required to facilitate mutual adjustment. It falls to government to articulate those regularities which facilitate social order. This being the case, lawmakers must be aware that their task is not the implementation of a particular pre-designated pattern of order, or the organization of individuals into a desired order, but rather it is their role to provide the conditions of relative epistemic certainty – stability of expectations – which allow individuals mutually to adjust and to create an order between them which will be spontaneous in form. Thus lawmakers intend that there be order, but their actions do not determine the precise nature of the order which results.

We should note at this point that Hayek relates this argument to his belief in the existence of concrete epistemic limitations on the ability of government to act efficiently. As we will see in the next chapter, he believes that it is impossible for a government to centralize the knowledge necessary to plan a social order in detail. This leads to the superior efficiency of self-adjustment and spontaneous order formation. The task of government is restricted by the 'knowledge problem', and its lawmaking function is shaped by a desire to make use of the spontaneous ordering devices that arise when individuals can rely on generalized rules of behaviour which provide stability of expectations. As we noted in the section on the origin of government the generalized rules are not rationally constructed by a government, but rather represent articulations of established opinion. Such is the importance of public opinion to the continued existence of any government that its actions will always be to a certain extent guided or circumscribed by it. If the law is to be effective in stabilizing expectations and inducing order, then it must be acceptable to the majority of the population. As Hayek notes: 'To become legitimized, the new rules have to obtain the approval of society at large – not by a formal vote, but by gradually spreading acceptance' (LLL vol. 3: 167). Government is limited as to the form that these generalized rules of conduct can take by the opinion of the people among whom they seek to induce order. One consequence of this is that the enactments of government ought to be restricted to areas where it is possible to secure an agreement of the majority (Hayek 1991: 45). This agreement need not be formally achieved, but if it were not at least possible then the enforcement of the general rule would require a degree of arbitrary coercive effort by the government which would destabilize expectations and prompt disorder.[15]

While we have noted that Hayek's intention is for government to provide a stable framework of general rules guided by the opinion of the people to be governed, we should also be aware that this is not by any means an inertial model. For Hayek the minimal state does not mean a state that does not act. He argues that when Smith spoke out against intervention by government he was referring only to actions by government which impinged on the protected spheres of individuals (Hayek 1960: 220). The enforcement of general rules does not represent interference by the government within the spontaneous order of society, but rather enhances the regularities necessary for the formation of a benign spontaneous order. The reform of those general rules, so long as it is carried out in the specified manner of couching the rules in general terms guided by opinion, does not represent interference in the order. Rather it should be considered as a process of refinement or immanent criticism. Such reform of the framework, properly undertaken, is a necessary part of the functional efficiency of the society as a whole. Hayek's metaphor for this is of a gardener: 'The attitude of the liberal towards society is like that of the gardener who tends a plant and in order to create the conditions most favourable to its growth must know as much as possible about its structure and the way it functions' (Hayek 1991: 14).

Our analysis thus far has stressed the non-deliberative and unintended nature of the development of social institutions and so the question remains as to where human agency and purposive rationality fit into the spontaneous order approach. It stands to reason that, though the rules which facilitate the formation of spontaneous orders are often the product of a process of evolution, there is no reason why this is necessarily always the case. Consciously 'made' rules in the correct generalized form might equally allow for spontaneous mutual adjustments. Moreover, spontaneously evolved rules are themselves subject to intentional reform. Part of Hayek's critique of constructivist rationalism was that this sort of intentional reform was being undertaken in a manner which risked destroying the benefits gained from evolved rules by ignoring the knowledge which is inherent in them. This, however, is not the same as saying that reform is impossible, or that reason is powerless to improve the efficiency of the framework of general rules.

This is precisely the line of argument advanced by Christina Petsoulas in her critique of Hayek's 'appropriation' of Mandeville, Hume and Smith. Petsoulas' argument depends, as do many critiques of Hayek, on an attempt to separate the 'twin' ideas of spontaneous order and evolution coupled with a critique of the latter. She does not take serious issue with mutual adjustment under general rules, but she does reject the notion of the non-deliberative evolution of the rules necessary for spontaneous order formation. She writes: 'Surely, rules which are deliberately altered, and which are maintained by intentional enforcement, cannot be the product of unconscious adaptation' (Petsoulas 2001: 5), and she notes that 'if evolved rules of just conduct have to be enforced by an external agent, such as the state, it cannot be claimed that the spontaneous order is self-maintaining' (Petsoulas 2001: 68). All of which leads her to conclude that: 'the end result is unforseen, but incremental intentional improvements on inherited traditions and innovations (designed or accidental) are *consciously* selected to survive because they are found to serve particular human goals' (Petsoulas 2001: 92). From our present analysis there is nothing in this last statement with which Hayek would disagree so long as it does not claim universality. As we have already noted, Hayek's idea of group selection in no way conflicts with his advocacy of immanent criticism as the preferred method of reform of social institutions. Indeed his rejection of conservatism might lead us to believe that successful reform by immanent criticism becomes a decisive factor in group selection as societies advance. Moreover, as has been argued through our conjectural history of morals and institutions, the rules which facilitate the spontaneous social order are not simply 'laws' in the obvious sense. Law is an articulation of opinion and opinion itself is a product of socialization into a particular moral tradition whose origins lie in mutual adaptation and conventional agreement. Law as an articulation of this is obviously on some level intentional. What Hayek tries to show is that the intent behind the articulation refers to some specific problem situation and not to the holistic aim of securing a particular social order: indeed such articulated laws as are enforced, are enforced by evolved institutions. For Petsoulas' argument to convince decisively against non-deliberative evolution she would have to show that government was intentional in its origins or that its reforms have always consciously been aimed at securing a particular systemic order rather than the resolution of particular disputes. In other words she would have to fall back on an approach which the Scots would criticize as a simple model or fail to offer an objection to Hayek's notion of immanent criticism. Government is an evolved institution grounded on opinion, that it acts purposively to secure stability of expectations is not the point of Hayek's argument. Rather it is that it did not arise purposively and is not capable of recreating the conditions, within which it did arise, in order consciously to reform itself in a holistic manner. The deliberate imposition and reform of rules of just conduct is not aimed at creating a new order, but at refining the existing order. Governments react to changes in opinion, and opinion represents the adjustment of individuals to their circumstances.

The paradigmatic example of this is not the market but the development of science that we discussed earlier. Science, for spontaneous order theorists, is a conscious attempt to mirror the process that constitutes the human mind. It is thus based on the higher order non-conscious 'rules' of the mind but is itself undertaken in a deliberate manner in an attempt to advance knowledge. Seen in this light the rules laid down by government as laws are attempts to improve the efficiency of the social order which are necessarily based on higher order non-deliberative rules such as morality, habit and tradition. Cultural evolution has advanced human understanding to a point where immanent criticism of traditional behaviour is eminently possible.[16] What Hayek and the modern liberals rail against is not reason but a conception of reason that sees it as existing outside the cultural tradition within which it has evolved. Gray is correct to understand this as: 'Hayek's attempt to combine respect for spontaneous traditional growths in law with the possibility of their rational assessment and critical selection' (Gray 1986: 71). Human social practices are the result of a process of cultural evolution but, like the state of debate in science, they are open to criticism from within the broad cultural tradition in which they exist. Politics ought to be undertaken in a manner akin to that advocated for science by Popper and Polanyi: open and free critical debate within a broadly accepted tradition leading to reforms aimed at improving the efficiency of the system drawn from evidence, supported by experience, in the form of social science. In other words the descriptive analysis of social institutions suggests an instrumental justification of freedom in order to facilitate mutual adjustment and reform.

We see that – while the original institutional framework of society, the rules which facilitate spontaneous orders, arise from a process of unintended consequences – it is possible for government to act to refine the rules which induce order. For Hayek it is the role of government to modify the rules which it inherits in reaction to changes in circumstances (Hayek 1978: 11). Such reform may be aimed at refining the stability of expectations in response to changed circumstances, or it may be to facilitate further the formation of order by better articulating conventional practice in order to eliminate potential sources of conflict.[17]

If the role of government is to facilitate spontaneous order by stabilizing expectations through the articulation of general rules drawn from the experience and opinion of the people, then any reforms which are carried out must be in line with established practice. A government ought to 'tinker' with the existing system of general rules and not seek utterly to redesign it. Reform, to borrow Popper's phrase, must be 'piecemeal' (Hayek 1960: 70) and be the refinement of established practice.[18] Government initiated reform of the framework builds on established rules and practices with the aim of removing 'gaps' or incoherence that restrict the functioning of the system. Thus understood the government is engaged in the task of reform as it 'consciously evolves' (Hayek 1991: 61) the rule of law. Reform is conducted within the context of an extant system or tradition of general rules and is pursued as the result of the experience of circumstances which reveal gaps in the system. Such reform is based on an internal evaluative judgement, the result of a process of immanent criticism which operates within the system of rules. Hayek notes that: 'we shall call "immanent criticism" this sort of criticism that moves within a given system of rules and judges particular rules in terms of their consistency or compatibility with all other recognized rules in inducing the formation of a certain kind of order of actions' (LLL vol. 2: 24).

Immanent criticism and reform based upon it are refinements of particular sections of the existing system rather than rejections of the system as a whole. This clearly links into Hayek's support for common law and judge-made law. Judges undertake the reform of the system of general rules by a process of immanent criticism designed to refine the stability of expectations by setting precedent. Such adjustment enhances the functioning of the system so long as it is carried out in the proper generalized form. The law-making function of government accordingly is the provision of a framework within which mutual adjustment can take place in an efficient manner. It represents a vital part of the invisible hand. It is, to repeat in conclusion, the inducement of spontaneous order by the provision of regularities, drawn from conventional opinion and practice, which encourages the efficient interaction of individuals.

The second core function of government is the provision of certain services which are deemed to be in the communal interest but which would not otherwise be provided if the community did not act collectively. Hayek follows Hume and Smith in accepting the existence of public goods that ought to be provided by government. There are cases where mutual adjustment under general rules does not provide a good in an efficient manner – usually, as Smith noted, because of a difficulty in controlling property rights and developing price mechanisms – and other means have to be found to secure its provision. The existence of such public goods is, for Hayek, 'unquestionable' (LLL vol. 3: 41), and their provision by government acts to supplement the market provision of other goods. It is the service function of government to provide such public goods from revenue raised by taxation. He lists a number of the services which fit into his conception of the proper scope of government action, and lays particular stress on those services which cannot by their nature be restricted to specific individuals: for example defence, local roads, land registers and maps.[19]

The notion of government service functions as supplementing the market leads Hayek to argue that there is no necessary reason why extensive welfare or social services should be incompatible with the operation of a market economy (Hayek 1991: 28). Indeed Hayek appears to take it for granted that some sort of welfare system is a feature of all governments. His support for this view appears to be based on some notion of moral humanitarianism, though he also notes that it could at a base level be justified by a fear of revolution and a desire to appease the poor (Hayek 1960: 285). In any case it is taken as given that there will be some form of welfare provision as part of the service functions of government. Such welfare, as a uniform minimum standard of living, exists 'outside' (Hayek 1991: 99) the market. What he means by this is that the aim of welfare legislation is to secure a minimum humanitarian standard of living available to all if required: a service which does not interfere with the operation of the market process. Such a welfare programme does not seek actively to redress any material inequality perceived to be the result of market interaction. It is not the result of a desire to create a state of social or distributive justice. Hayek's conception of welfare 'outside the market' is a notion of welfare as a safety-net, a guaranteed minimum level below which no one is allowed to fall. Such welfare is concerned primarily with the relief of poverty and not with the redistribution of income. Which is to say that the service provision has no aim in view – such as a desire for greater equality – other than the relief of the difficulties of those whose incomes fall below a certain level. Such a model of welfare operates in accordance with generalized rules or principles, it treats all who have need of it in the same way, as opposed to a redistributive model that necessarily treats individuals in an arbitrary manner in order to equalize inequalities. Set incomes are not guaranteed to groups within the market, but only to anyone who should fall 'below' or 'out of' the market.[20]

His essential argument is that welfare provision is a feature of all developed societies, but that the development is necessary for it to arise. If welfare is to be viewed as a public good to be funded by taxation then it stands to reason that the wealthier the community, the higher the tax yield and the greater the provision of welfare. Under this view the market creates wealth and that wealth, through taxation, pays for a welfare system which exists 'outside' the market. We need wealth, and consequently the market order which creates it, if we are to provide improved welfare services (LLL vol. 3: 55). The improvement of welfare provision is an unintended by-product of the success of the market in generating wealth.

The service and welfare functions of government are carried out within a framework of general rules which exists 'outside' the market. Hayek notes the following limitations which this places on the service functions of government:

> 1) government does not claim a monopoly and new methods of render-ing services through the market (for example, in some now covered by social insurance) are not prevented; 2) the means are raised by taxation on uniform principles and taxation is not used as an instrument for the redistribution of income; and, 3) the wants satisfied are collective wants of the community as a whole not merely collective wants of particular groups.
>
> (Hayek 1978: 111)

Two aspects of Hayek's development of these points are of interest. First, and linked to our earlier discussion of the importance of the utilization of local knowledge, is the focus on devolved or localized administration of public services (Hayek 1978: 144–5). The use of local knowledge to apply resources is essential in Hayek's view because of the epistemic constraints placed on government by the dispersed nature of knowledge. There is no reason for, say public parks, to be administered by central government (Hayek 1960: 375). The efficient use of local knowledge is more likely to be obtained if decisions are taken as close as possible to the objects concerned. Second, the decision to keep control on a local level allows not just more efficient adjustment to circumstances, but also innovation, evolution and diversity among the means chosen to administer local government services (Hayek 1960: 261–3, 286). Competition between devolved authorities, emulation and experimentation, create an atmosphere of innovation which encourages the more efficient provision of service functions (Hayek 1978: 112).

Once again we see the key features of the spontaneous order approach applied through a conjectural history of the human institutions of property, law and government. The desire for order and stability of expectations that arises from humanity's nature as classificatory beings leads to the development of interpersonal conventions which ease social interaction. The desire to resolve disputes over conventions of property leads to the emergence of judges/chiefs whose articulations of customary practice in reaction to given problem situations is the origin of law and who themselves represent the emergence of government. This process evolves and adapts in reaction to circumstances leading to the development of the bi-functionary modern conception of government. Behind all this, as always, is an invisible hand argument about the epistemological efficiency of a framework of general rules that allows freedom for mutual adjustment and reform.

9 The evolution of markets

The division of knowledge

Throughout the body of his economic writings it is clear that Hayek is pre-occupied with the significance of the role of knowledge in the economy.[1] Drawing on the arguments of the socialist calculation debates of the inter-war years, Hayek notes that the successful functioning of a large society depends on the successful co-ordination of the knowledge held by millions of individual minds. The sheer volume and complexity of this knowledge is such that no one mind has the power to absorb it in its totality. No one human mind can possibly grasp all of the relevant data necessary to plan a whole economic system. Indeed it is an error (what Hayek calls the synoptic delusion) to proceed as if such a thing were possible. And this is Hayek's chief complaint against his opponents in the socialist calculation debate: their assumption that a rational plan for the whole of the economy is a matter of rational study of the relevant data. The central point about eco-nomics is that the information necessary cannot be centralized; indeed, the study of economics is the study of how individuals without such perfect knowledge interact. Hayek and his fellow Austrians stress the key role of the information-carrying function of prices and the discovery or information-finding function of entrepreneurs. They argue that these roles are missing from a system of socialist or command economics, leading socialist econo-mists to the logically impossible assertion that one mind or group of minds is capable of controlling and acting on all the information previously held by millions of individuals. The argument is essentially that the complexity of social phenomena prevents efficient economic planning and it runs, in an abbreviated form, like this.

All human minds are finite, and all economies are vastly complex: no one mind is capable of absorbing all of the information necessary to direct it suc-cessfully. Even if a super-mind were to exist the rapidly changing nature of circumstances and tastes would render any plan formed by such a mind obsolete almost as soon as it was made and therefore a less efficient ordering principle than mutual adaptation. Even should such a mind find a way to centralize and adapt, through some super-computer – though the number of equations requiring to be solved simultaneously suggest that this too is impossible (Hayek 1984: 58–9) – it is then faced by the problem that some of the knowledge used by individuals to guide their economic decisions is of a tacit, inexplicable, habitual nature that cannot be rendered in a form which reconciles it with the terms of the plan.

Human society is highly complex and, as we noted before, the chief subject matter of social science for Hayek is the study of the unintended consequences of human action: so the study of society is an examination of the 'phenomena of unorganized complexity' (Hayek 1984: 269), of the spontaneous order of society. Now it is already evident that it is constitutionally difficult to explain any very complex phenomena. The scope of the human mind is limited and it is an empirical truth that such minds are incapable of absorbing all of the relevant features of a complex situation (Hayek 1991: 44; 1960: 4). The mind cannot ascertain all of the data involved in the vast web of social interaction that forms the complex phenomena which make up society. And, as we noted before, such complexity prevents the detailed prediction of future events in specific detail (a key feature of historicism). The reduction of social phenomena to simple propositions, or historical laws, which would allow such prediction is impossible. There are no 'scientific' or eternal laws of society that can be ascertained by humans and used to assist in the future planning of society. Science can assist us better to understand the general nature of social inter-relationships, but it cannot ever hope to provide formulae that allow the accurate prediction of all future human circumstances. For this reason those who seek to plan an economy tend to pursue one of two courses. They either stick with what they 'know', that is to say they act in such a limited manner that their predictions have a reasonable chance of success (Hayek 1967: 80); in which case they plan very little or in such broad terms as to be highly inefficient. Or they seek to restrict the complexity of the circumstances in order better to understand and control them.[2] As Hayek notes 'made' or deliberately designed orders of this type are necessarily simple because our minds are capable of only simple ordering processes on a conscious level (LLL vol. 1: 38). Thus by limiting the knowledge necessary to understand an order we simplify the order and restrict its efficiency.

The complexity of social phenomena means that the analysis of them is very difficult; but it also means that the rational control and organization of them is a near impossibility. That it is not a strict impossibility is because it is possible to organize simple social interactions, but such organization would entail an abandonment of complexity and its attendant benefits. As the complexity of human knowledge and interaction grows we see an increasing process of specialization, a process that both compounds social complexity and is an expression of it. Hayek's essential point is that the complexity of interaction and the multitude of circumstances which are features of human society necessarily lead to a situation where they are more efficiently governed by individual adjustment rather than by planning or central control. Mutual adjustment, and the mechanisms which facilitate it, allow individuals to benefit from the knowledge of others by following certain signals which communicate more knowledge than could possibly be assimilated and analysed by any one mind.

Hayek follows the Austrian practice of rejecting classical equilibrium analysis in economics (Hayek 1980: 35).[3] He believes that such an approach is unrealistic on account of its assumption of perfect competition arising from perfect knowledge. For Hayek, the instrumental justification of a market economy does not arise from any conceptual model positing the efficiency of perfect competition. This is because basing an argument in favour of the efficiency of market competition on a model of perfect competition entails a conceptual model that posits perfect knowledge on the part of the participants. Hayek argues that such perfect knowledge is impossible and, that even if a supposition of it is to guide a conceptual model, it could not be conceived in a manner that would make the model applicable to any real life situation. To believe in models of perfect knowledge is, for Hayek, to misunderstand the nature of the economic problem which is fundamentally one of the imperfect nature of human knowledge. The value of competition for Hayek, and the Austrians, is precisely that it is conducted in a situation of imperfect knowledge, and that it acts as a discovery procedure which coordinates various imperfect collections of knowledge (Hayek 1980: 96).

As we noted above, Hayek argues that it is because all of the relevant information is not available to one mind that planning fails and competition is a more efficient economic model. Hayek argues that the human mind is imperfect and, as a result, all individually held human knowledge is imperfect. As we saw in Chapter 6 on science, all human knowledge is essentially a system of categorization that proceeds in a negative manner. Which is to say that it functions by excluding possibilities rather than by making definite eternal assertions. To repeat the earlier point, the growth of knowledge is not the growth of certainty but the reduction of uncertainty. We also noted, in Popper's critique of holism in particular, that the limited nature of the human intellect means that we cannot ever hope to conceive or grasp the 'whole' situation in all its aspects: which means that we cannot, by the nature of our minds, know everything about everything. Our minds are restricted and act efficiently only when focused on a particular problem, which we, to a degree, isolate from its surrounding phenomena. As a result of the limited nature of our minds, these focused studies are necessarily partial studies of elements of the 'whole' of social phenomena (Hayek 1967: 124). This leads Hayek to the assertion that humans are simply not clever enough to have designed an economic or social system. No one mind can grasp all of the facts necessary to create such a design and no tool of human understanding could ever make such a design possible.

For Hayek this marks the vital 'importance of our ignorance' (Hayek 1967: 39). Humans are necessarily ignorant of most of the facts of social interaction. This ignorance leads him to assert that the most important task of reason is to realize its own limitations. To realize the limits of our rational abilities, and those of our knowledge, is to follow the Socratic approach to philosophy (Hayek 1960: 22). It is to develop a 'humble' (Hayek 1960: 8) view of human reason and knowledge that recognizes the limited nature of our capacity to understand.[4] If we recognize the limits of our individual and cumulative knowledge we are, Hayek argues, in a better position to make effective use of that limited stock which we do possess. As he argues: 'I confess that I prefer true but imperfect knowledge, even if it leaves much indetermined and unpredictable, to a pretence of exact knowledge that is likely to be false' (Hayek 1984: 272).[5]

Hayek's argument is that our knowledge is always imperfect and dispersed among many individuals. As individuals we cannot know all of the consequences of an action nor can we know all of the circumstances under which it is carried out. We must necessarily act on what knowledge we have but also, and this is crucial, we are enabled to act on information held by others *without explicitly knowing it* through the development of certain social institutions which facilitate the transmission of knowledge in an abbreviated form. Such a process does not depend on any centralization of all knowledge, rather it functions by allowing individual adjustment to the behaviour of others who are similarly adjusting to their circumstances. We have developed these social practices and institutions without any real idea of their overall nature or operations (Hayek 1978: 71). That is to say an institution such as the market allows us to act as if we have that knowledge held by others even though we do not. Moreover, we are not even aware of the totality of the system that allows us to act in such a manner. Such institutions (price, propriety, law) are adaptations to our ignorance that function to reduce the necessary uncertainty which arises from our ignorance of the knowledge held by others.

Whilst human knowledge does indeed exist, and may even be considered to form a cumulative whole, this whole does not exist in a unitary form nor is it within the grasp of any individual.[6] Human knowledge is dispersed among 'countless' (Hayek 1984: 270) people, it is held by individuals and is not capable of unification in any complete sense. Individuals are the media of knowledge in this sense: they are the ones that 'know'. They cannot have knowledge of the whole but they may have knowledge of many discrete individual facts. It is in this sense that our knowledge of the knowledge of others is similarly limited. For example, under the division of labour, workers in distinct specialities may have no idea of the knowledge upon which they depend: the man who puts the heads on pins need have no knowledge of the production of the steel upon which he depends for his materials. Ignorance is mitigated by interdependence, we need not know the knowledge of others in order to benefit from it. Hayek follows Smith in regarding this as one of the chief benefits of a system of trade and market exchange. We are able to specialize and to exchange, to profit and to survive, without the need for extensive knowledge outside our chosen specialization.

The market does indeed co-ordinate the knowledge of many millions of individuals, but it does so without collecting that knowledge in one place at one time. The interaction of individuals, as the media of trade, is the media for the exchange and interaction of information. It allows dispersed knowledge, the knowledge of different circumstances held by different individuals, to be of use to all people.

Hayek follows Smith in noting the spontaneous, undirected nature of the division of labour, and in finding it to be constitutive of a complex society (LLL vol. 1: 110). We saw in Chapter 5 how specialization and the utilization of dispersed knowledge are vital to Smith's understanding of the benefits of the division of labour: but Hayek does more to stress the role of knowledge in this process than his predecessor. He states that the benefit of the division of labour is that it is based on a complementary division of knowledge. Not only does the division of knowledge, produced by the division of labour, lead to material gains, but it also (as a 'fact' of economic 'reality') places constraints on what individuals can hope to do consciously to develop the economy as a whole. The increasing importance of specialist knowledge as the cumulative sum of human knowledge advances, the increasing likelihood that a specialist will know more about an aspect of human knowledge yet less about many others, lies at the heart of economic development. For example, as the cumulative sum of academic knowledge advances, so Hayek believes, specializations will become increasingly focused (Hayek 1967: 123–5). The sum grows not because individuals know more in the sense of encompassing wider fields, but rather because the depth of their knowledge of a narrower field adds to the whole: the moral philosopher of Smith's day has been succeeded not only by the modern philosopher but also by the economist, the political scientist, the anthropologist and a host of other specialists.

The different experiences of diverse individuals broaden human knowledge. For example, the development of a profession proceeds by the development of those who profess it. This is, as Hume noted, not to say that humans are becoming in some sense more intelligent in the individual sense of mental capacity. Rather it is to say that the focusing of attention necessarily increases the scope for useful observation by individuals of necessarily limited mental faculties. The division of knowledge is both a reaction to the complexity of the world and a process which compounds that complexity by diversifying the experience and knowledge of individual specialists. The growth of cumulative social knowledge is a product of the division of knowledge achieved through specialization. While at the very same time this growth of cumulative knowledge encourages and requires further divisions of knowledge in order to deal with the increasing scope of human inquiry and behaviour.

As we have seen, a complex, modern, open society depends on a division of knowledge that precludes large-scale social planning. The atavism that Hayek associates with constructivist rationalism, the desire to return to some version of a tribal society, links into this argument. It would be possible, in Hayek's view, to plan or organize a small-scale tribal society with a tolerable degree of success. But this is simply because the amount of information required, like the number of individuals involved, would be strictly limited (LLL vol. 2: 88–9). Once population grows and societies reach even moderate degrees of size and complexity, the task of efficiently planning economic activity exceeds the capabilities of any one mind. Moreover, as we shall see, it is only because of the division of labour and the division of knowledge that population growth of this sort is possible at all.

Localized, individually held knowledge cannot accurately be centralized. Such local knowledge is knowledge of particular times and circumstances held by those who experience them. It is acquired from the specific position of the individuals concerned who shape their individual plans as an adaptation to those circumstances. Individuals within a particular set of circumstances will use their knowledge of those circumstances to select the aims of their actions. For Hayek this is simply an extension of the division of labour argument about specialization: knowledge of the circumstances of one's profession places one in a position to make the most effective decision regarding the objects concerned. If such knowledge of circumstances is individually held, then the most efficient overall use of knowledge will be produced by the cumulative sum of the successful exploitation of local circumstances by individuals (Hayek 1978: 133). This efficiency is further underlined if the case of a change in local circumstances is considered. The efficient utilization of human knowledge requires the successful adaptation to changes in local circumstances, and successful adaptation is once again most likely to be produced by those with the relevant first-hand knowledge (LLL vol. 2: 121). Individuals form plans in reaction to the circumstances in which they find themselves, and these plans are adapted to changes within those circumstances as they occur. Individual exploitation of local circumstances and adjustment to change depends upon the successful exploitation of such localized knowledge. If, as Hayek argues, such knowledge is held individually and cannot successfully be centralized, then it follows that a decentralized system, one which depends on the adjustment to circumstances by individuals in those circumstances, is the most efficient system for the adaptation to change.[7]

It might be countered that some form of decentralized planning might be utilized to get around this aspect of the knowledge problem but this, as Hayek notes, neglects the fact that it is not merely the possession of localized knowledge which is significant but also its use – use in the sense of its integration with other localized knowledge. It is only one part of the argument to say that the 'man on the spot' (Hayek 1984: 217) is in the best position to make effective decisions about the particular circumstances in which he finds himself. The effective management of resources by placing decision-making in the hands of those on the ground is an important part of Hayek's argument about the efficient use of knowledge.[8] However, the question remains as to how best to encourage the co-ordination of such knowledge while preserving a degree of flexibility and autonomy which would allow the benefits of de-centralization to be enjoyed. What Hayek seeks to understand is how efficient use can be made of localized knowledge in such a way that its benefits are transmitted throughout society in an effective manner. He wishes to understand how individuals adjusting to changes in local circumstances are able to communicate this knowledge in such a way that other individuals in other circumstances are able to adjust to their adjustments.

Taking this view of human knowledge as the imperfect, diverse and dispersed reactions of individuals to the circumstances in which they find themselves, the question becomes one of how humans successfully reconcile these adaptations and sums of local knowledge: it becomes the problem of the co-ordination of knowledge. Here Hayek draws on his previous thoughts on human psychology: chiefly the desire for order and the dislike of disorder. Hayek argues that the human propensity to generalize from experience leads humans to form 'rules' (LLL vol. 2: 23) of adaptation to particular kinds of recurrent circumstance. Through a process of socialization these rules of reaction are spread through society. As a result individuals need not share local knowledge of all relevant circumstances so long as they proceed by such general rules of adaptive behaviour. If such interaction does not depend on an explicit situational analysis, but rather on the following of rules which stabilize expectations, then there is no reason why such successfully socialized conventions of behaviour need be consciously understood and undertaken in order to fulfil their function. They may instead exist as customs, habits or traditions of behaviour whose successful functioning is not dependent on deliberative understanding. Hayek proceeds to examine the human institutions that have arisen to facilitate the successful adaptation of individuals to circumstances of which they are unaware. General rules of behaviour provide a degree of stability of expectation, but what must also be taken into account is how humans have developed co-ordination devices which allow successful mutual adjustment under those rules.

The co-ordination of knowledge

As we have seen, Hayek considers human knowledge to exist in the dispersed, imperfect holdings of individuals. Following on from his analysis of the division of knowledge he asserts that civilization rests on the effective use of this dispersed knowledge. Put another way, civilization rests on the co-ordination of dispersed knowledge efficiently used: upon the invisible hand. As Hayek notes, this approach suggests that civilization does not so much depend on a growth of knowledge as on the efficient use of knowledge already held (LLL vol. 1: 14). The question is how, or more precisely by way of what institutions, do we integrate this dispersed knowledge given that it is impossible to centralize in one mind. This question, the co-ordination of dispersed knowledge, is the central economic problem (Hayek 1980: 50–4).

As we noted when we discussed central planning there are epistemic limitations placed upon the scope of a central authority's ability to act in an efficient manner in this regard. Humans, however, in Hayek's view, have developed a series of institutions and practices (the invisible hand), which he terms the 'extended order' (Hayek 1988: 81), which assist us in dealing with our ignorance. The efficient co-ordination of dispersed knowledge is necessarily an 'inter-individual process' (Hayek 1979: 152). There are two issues at stake: how to make use of the individual sums of knowledge, and how to co-ordinate the use of unique local knowledge of circumstances. In terms of economics the development of a system of price signals operating within an extended market order – or 'Catallaxy' (Hayek 1984: 258) – is the key human practice that functions to deal with our ignorance and co-ordinate dispersed knowledge (Hayek 1980: 17).[9] The market produces a co-ordination of knowledge by facilitating mutual adjustment in reaction to the information held by prices: 'It is through the mutually adjusted efforts of many people that more knowledge is utilized than any one individual possesses or than it is possible to synthesize intellectually; and it is through such utilization of dispersed knowledge that achievements are made possible greater than any single mind can foresee' (Hayek 1960: 30–1). A market system with prices and their incentive qualities direct attention, facilitate mutual adaptation and function to encourage the efficient utilization of dispersed knowledge (Hayek 1984: 326).[10] The efficient use of individual knowledge by mutual adjustment is the true nature of what Smith metaphorically termed the invisible hand. The invisible hand produces benign results precisely because it ensures the efficient utilization of knowledge to produce the best possible results given the limited and dispersed nature of individual sums of knowledge. The invisible hand is the institutions and practices that allow the successful co-ordination of human knowledge and effort through mutual adjustment.

In an attempt to make use of dispersed knowledge in a complex extended society the dictates of efficiency lead us to favour decentralization. Such decentralization requires some institution which co-ordinates the decentralized decisions, some process of mutual adjustment that ensures that decentralized decisions might accommodate themselves to each other. As we have seen, to trust this role to a central planning authority, or to one mind, limits the adaptability of the decentralized decision-makers and is inefficient as a result of the epistemic limitations of the human mind. Competition for Hayek is the practice that humanity has developed in an economic context to deal with this problem. However, as we noted above Hayek is not concerned with static, unrealistic models of perfect competition. What he seeks to understand is actual competition, viewing the market as justified precisely because competition, however imperfect, is preferable in terms of efficiency to any other mode of co-ordination.

Hayek argues that competition fosters a spirit of 'experimentation' (Hayek 1960: 261) which encourages us to seek after new ways of doing things. It is only, he argues, through a competitive process that we discover facts and exploit new, more efficient, modes of behaviour. He writes of the relationship between competition and knowledge that:

> Competition is essentially a process of the formation of opinion: by spreading information, it creates that unity and coherence of the economic system which we presuppose when we think of it as one market. It creates the views people have about what is best and cheapest, and it is because of it that people know at least as much about possibilities and opportunities as they in fact do.
>
> (Hayek 1980: 106)

Competition is a 'discovery procedure' (Hayek 1988: 19), an adaptation to circumstances that promotes the efficient utilization and discovery of knowledge. The competitive process of a market economy is the medium for the acquisition and communication of knowledge. Competition, the desire to succeed in such, is the incentive that moves us to seek out new ways of doing things; and our observation of the success of others in this process is the transfer device for efficient utilization of knowledge. The market is primarily to be understood as a communication system with competition as the device that keeps the messages carried as accurate as possible.

While Hayek's most significant discussion of the co-ordination of knowledge is to be found in his analysis of the functioning of a market economy, we should note from the start that he does not consider the 'market' to be a single edifice or institution that may be analysed as a whole (Hayek 1984: 219). The term market now no longer applies to buying and selling, to trade, in a specific location which may be termed a marketplace. Rather it may be understood as a form of spontaneous order, as a mode of human interaction which is most closely akin to what is commonly understood by the term economy. However, Hayek dislikes this term, believing that the spontaneous market order, or catallaxy, is formed from the interaction of a broad range of economies – in the Aristotelian sense of household management (LLL vol. 2: 108–9). The basic human desire to provide for subsistence, arising from the 'natural' biogenic drives of the body, is the force which has driven the creation and development of the market order. A series of institutions, practices and interrelations has developed from experience which allow the efficient exploitation of dispersed knowledge for mutual advantage. By way of the institutions of the catallaxy the total product of human labour is maximized because of the maximization of the product of individual human labour (LLL vol. 2: 118). The market increases an individual's ability to adapt successfully to his circumstances, while simultaneously promoting the more efficient use of resources. In addition to the efficiency and adaptive characteristics of the market order which promote the maximization of production, the market also acts in a social or cultural manner such that it diffuses power through society (LLL vol. 2: 99), as the

Scots noted in their analysis of the fall of feudalism, and also such that it promotes the peaceful interaction of peoples (LLL vol. 2: 113).

At the heart of Hayek's analysis of the catallaxy is his understanding of its role in terms of information and knowledge. He views the market as an impersonal device which directs individuals toward profitable behaviour.[11] The system of market prices acts as an 'impersonal mechanism' for information transmission (Hayek 1991: 37); as a 'mechanism for communicating information' (Hayek 1984: 219); a 'steering mechanism' (Hayek 1960: 282) and moreover as a communication 'system' (Hayek 1984: 276). Drawing on the analysis of the Austrian tradition of economics Hayek understands prices as signals that 'abridge' (Hayek 1979: 173) information, which render it into a comprehensible form that may be used by individuals with a limited capacity to absorb information. The trial and error process of bargaining through which prices are established filters information about human desires and resources and creates a monetary 'value' that guides the action of individuals in relation to productive activity. As a result a market, through the price mechanism, spreads information in a comprehensible form which may be used by individuals in the formation of those plans which guide their action.

Prices are signals: they direct the attention of individuals towards significant areas of concern without requiring those individuals to comprehend all of the information which they embody.[12] Such signals allow individuals to adjust their behaviour in order to pursue the most profitable course of action for themselves: they embody information about profitable behaviour, about the demand for and scarcity of goods, which allows individuals to adapt to circumstances beyond their individual experience. For example, prices, in the form of wages, indicate the demand for certain professions. They direct the form of specialization or the division of labour and in a sense are necessary for such specialization to occur in an efficient manner. However, as Hayek notes, this directive function of prices in relation to specialization has little to do with notions of reward or desert. The function of prices is not to reward people for work in the past, but rather it is to indicate profitable occupations or specialities to others. Prices do not reward people in this functional sense; they direct them toward profitable modes of behaviour by indicating demand (LLL vol. 2: 72). Prices act as incentives not only to those already within the speciality, but also to those outside it. Price signals flag up incentives, they indicate areas of profitable specialization. This is what is meant by the phrase 'to spot a gap in the market'. An entrepreneur will exploit a 'gap' in provision indicated by price signals and his success, or profit, will act as an incentive to others to join him in his profitable specialization (Hayek 1984: 219, 262).

Moreover, because price is a subjective value, in the sense that it is interpersonally generated through the interaction of many buyers and sellers in a diversity of circumstances, it is open to frequent fluctuation and adaptation to changes in those circumstances which generate it (LLL vol. 3: 170).[13] A change in price is an adaptation to a change in circumstance that 'abridges' the knowledge of that change and renders it understandable to individuals without explicit knowledge of those circumstances. The inter-subjective nature of prices is what allows their quick and accurate alteration to reflect the knowledge of changed circumstances held by individuals. Up-to-date prices reflect up-to-date information in a form which cannot efficiently be procured by a central planning authority. Which is to say not only can the authority not absorb that information, but also it cannot adapt quickly enough to gather new information on changes in circumstances. Prices are an evolved institution that has developed as a response to our individual ignorance or limited information-processing abilities. They allow individuals, through the 'reading' of them, to adapt to changes in circumstances beyond their experience and to alter their plans accordingly.

The market order works through a system of incentives which embody information and direct human activity. For these prices to carry accurate information about profitable activity there must be room for the disincentive of failure. Signals about unprofitable behaviour are equally as important, in a directive function, as signals about gaps in the market and opportunities for profit. A competitive process underlines this function of prices with competition between actors securing the efficient allocation of resources. Inefficient producers will fail and the failure will serve as a signal of un-profitability (Hayek 1980: 176). Hayek regards this function of the market order, and its expression in the form of prices, as a form of negative feedback. Information about unproductive behaviour is relayed to other individuals by the failure of those who attempt it. Such negative feedback requires that frustration and failure should be allowed to occur in order for the information to pass on to others. It is here that we see the beginnings of Hayek's epistemological argument for freedom in the market order. The freedom to fail must exist just as the freedom to exploit local circumstances must: in order for the information processing of the price system to operate efficiently.

By this understanding freedom is justified as a means to progress. The problem remains that Hayek provides no argument in support of freedom as a value in itself, nor indeed does he provide a consistent definition of the concept, a point which rankles some of his critics.[14] Though, as we have noted throughout this study, this criticism does not refer to the internal consistency of Hayek's application of the spontaneous order approach, but rather to what would be required to supplement it should it be analysed in moral terms.

As we noted above Hayek believed that epistemological constraints render calls for central planning of economic activity unrealistic. In a command economy there can be no accurate transferral of information through freely adjusting prices. The question that is central to all attempts to control an economy is how do we arrive at a measure of value which accurately reflects the circumstances in the absence of a price system? With

Mises, Hayek argues that this problem is insoluble: the role of prices as information signals and subjective standards of value cannot be replicated by any deliberately designed system. The failure of planned economies is caused by their inability to utilize the knowledge held in prices. Indeed efficient economic co-ordination in a market order is only possible because of prices, and their absence in any economic system would make effective co-ordination impossible. Moreover, Hayek also attacks the more interventionist economic management techniques of those Western governments influenced by the economic thought of Keynes. He argues that any attempt to control prices or wages acts as a perversion of information transmission. The information carried no longer reflects in an accurate manner the circumstances of production, but becomes infected with the policy desires of governments (LLL vol. 2: 76). Price controls fail in Hayek's view because they misdirect economic actors, they lose the efficient co-ordination of inter-subjective market prices. Moreover, such controls require a degree of arbitrary interference by government, and this has a disruptive effect on the whole of the catallaxy, reducing its successful operation by hindering the accuracy of adaptation. Hayek argues:

> Any attempt to control prices or quantities of particular commodities deprives competition of its power of bringing about an effective co-ordination of individual efforts, because price changes then cease to register all the relevant changes in circumstances and no longer provide a reliable guide to the individual's actions.
>
> (Hayek 1991: 27)[15]

The desire to ensure 'just' remuneration which lies behind price and wage controls is a mistaken approach. Following Menger he stresses the view that prices embody a subjective valuation: they do not, and cannot meaningfully be understood as, expressing an objective value such as desert (Hayek 1984: 199).[16] Drawing on the Austrian theory of value Hayek argues that prices are generated interpersonally and as a result are subjective. They do not embody any objective or universal value such as desert.[17] Money, for Hayek, is an institution that humans have developed to ease the process of market interaction and which functions to provide abridgements of information. It is a means and not an end. Moreover, because the market deals with interpersonal, subjective valuations in monetary terms it makes no sense to talk of value as inhering in concepts such as labour (Hayek 1984: 145). The spontaneous adjustments of the market order are efficient, but they are not moral (LLL vol. 2: 63), they do not deal with concepts such as desert.

The catallaxy succeeds by being a game played under general rules. Which is to say that fairness in this sense applies to the way the game is played and not to its results. For Hayek it makes no sense to view the results of market interaction in terms of moral evaluations because the market is both non-moral and non-purposive (LLL vol. 2: 62). Since the results of market interaction do not represent the product of any one actor, or actors, action and intentions; then it is impossible to judge its results with reference to moral assertions concerning desert. In Hayek's view the market is both neutral and 'blind', it does not seek to maximize results for all but merely provides the best chance for those results to occur by allowing the accurate transmission of information. Moreover, and significantly, the market is not a zero sum game. There is no cake that is divided as a prize in the sense that the very playing of the game in an efficient manner increases the size of the potential prize (Hayek 1984: 260; LLL vol. 2: 115).

In conclusion, the application of the spontaneous order approach shows that the market represents a system of mutual adjustment providing for the efficient use of knowledge within an evolved set of social institutions and practices. The limited nature of human knowledge leads to the development of adjustment mechanisms that allow the efficient use of dispersed and imperfect knowledge within a framework of general rules which provide a degree of stability of expectations. Once again freedom, bounded by general rules, is held to be instrumentally justified to encourage the accurate operation of mutual adjustment. In this study of the role of prices we find the nature of the invisible hand. It is the evolved institutions and practices that allow the efficient discovery, communication and utilization of knowledge.

10 The invisible hand

We have now reached a point where we can draw together the various elements that typify the spontaneous order approach to political theory. Throughout our examination of the Scots' and the Moderns' discussions of the nature and origins of the core social institutions we have seen how they deploy a series of concepts that constitute a particular explanatory approach to the study of social theory. Spontaneous order theorists believe that they are engaged in a descriptive, scientific project that aims at an accurate understanding of the social world. The spontaneous order liberals set out to explain the nature of the social world, and from that explanation they seek to draw conclusions about the most effective means of securing what they believe to be a group of universal human goals. These goals are not laid down in the language of moral values, but rather are drawn from an examination of the factors that universally motivate human action. Thus, the spontaneous order theorists argue, all humans seek to secure subsistence and material comfort. As a result social systems that secure these goals can be regarded as successful.

Spontaneous order thus represents a distinctive approach to social and political theory within the liberal tradition. As we noted in the first chapter it exists within a strain of Anglo-American empiricist thought that we termed British Whig Evolutionary Liberalism. We have seen how this evolutionary approach, in particular its stress on habit and non-deliberative behaviour, distinguishes the spontaneous order liberals from the more rationalist approach of 'continental' liberals such as Kant and the Benthamite utilitarians. It can also be contrasted with both the justificatory contract approach of liberal theorists such as John Rawls, and the rights-based approach of libertarians such as Robert Nozick. Moreover, the heart of the difference lies not in the principles defended or the conclusions drawn, but in the method of argument. Spontaneous order theorists rest their argument on a descriptive social theory rather than a normative moral argument in favour of liberal principles. It is therefore important to bear this distinction in mind when examining the work of the Scots and the Moderns. This is because their mode of argumentation differs from that of the other liberals: their approach and the evidence that they cite in support of their position is far more rooted in a descriptive social theory rather than a prescriptive moral argument. Liberalism, for both the Scots and the Moderns, is instrumentally justified as a result of the scientific observation of social phenomena: it is concrete rather than abstract, its conclusions rest on a series of falsifiable assertions about the efficiency of liberal institutions rather than an abstract vision of the Good. Through our focus we have been able to demonstrate this aspect of the Scots and the Modern's argument. Moreover, by examining their writings on the same linked topics we have seen how the approach consistently deploys the same core concepts as it seeks explanation in each field.

Spontaneous order theorists of both periods operate with a particular conception of human nature or the underlying universalities of human behaviour. They argue that human beings are sociable creatures who can only be understood within the context of a social setting. A psychology of human behaviour is developed that views the human mind as an ordering device that 'sorts' our experience in order to calm our minds and to stabilize our expectations. Humans are by nature order-seeking beings who learn from their experience of the world around them, and who adjust their behaviour, and their mental order, in reaction to their circumstances. The chief underlying characteristic of human behaviour is the desire to provide for subsistence. This is a universal human aim drawn from the biogenic drives that constitute our animal nature, and can be used as an underlying universal principle through which to analyse the development of social institutions.

Spontaneous order theorists like Adam Smith believe that they are engaged in a 'scientific' project. They regard science as the search for order, and the classification of experience, to 'explain' that which appears wondrous to us. Science represents a formalized, deliberative version of the operation of the human mind undertaken to stabilize our expectations. Social science is the explanation of order in the social world: it is conducted by the examination of historical evidence and the formation of composite models, or conjectural histories, of the institutions and practices that constitute the social order. Both the Scots and the Moderns regard the origin of the core social institutions – science, morality, law, government and the market – as traceable to the interaction of order-seeking individuals with the circumstances in which they find themselves. However, these institutions are not originally the deliberate result of the purposive actions of individuals. Mankind did not set out rationally to construct these institutions. They sought to stabilize their expectations and produced, as a result of their interaction, a series of unintended consequences that led to the formation of social order. Such a spontaneous order approach remains, at base, a methodological individualist one: the explanation still invokes the behaviour of individuals as the primary unit of understanding, but insists that the order which they form was not part of the intentions behind their original action. Social order is spontaneous, the result of the mutual adjustment of individuals to their circumstances and to each other. The origin of social institutions and practices does not lie in deliberate design and both groups of thinkers reject simple models or constructivist rationalist approaches to social theory. Indeed this approach regards reason itself as a product of the spontaneous process of the development of order: social institutions could not be the product of deliberative rationality because deliberative rationality could only have been developed in a social context characterized by the order and stability produced by those institutions.

Social institutions originate as conventions, at the level of group rather than individual, habit formation. These conventions represent an inter-subjective equilibrium reached by the actors which, from experience, they have found effective in stabilizing their interactions. This stabilization allows individuals to pursue their own purposes more effectually by reducing uncertainty as regards the actions of others. The social order evolves and adjusts to changes in circumstances as they occur so as to preserve stability. This evolution is often, but not necessarily, non-deliberative: though the end of our adjustments is not a specific pattern of order it is possible for us to apply our knowledge and experience to enhance the order-inducing characteristics of our institutions. Indeed social science that proceeds by the spontaneous order approach provides us with a degree of understanding that might form the basis for a process of immanent criticism and reform of existing institutions.

At this point the spontaneous order theorists develop a second line of argument that we have termed, following Smith, an invisible hand argument. Given that society is properly understood as an evolving spontaneous order, that is the product of the unintended consequences of the actions of subsistence and order-seeking individuals, then an explanation is required to account for the benign, or successfully functioning, nature of particular spontaneous institutional orders. Drawing on the underlying universal human goals of order and material comfort, they regard it as clear that some institutional adaptations have met with more success than others. What the spontaneous order theorists set out to examine is what features of these orders produce this success, and how they might be extended to provide the best chance for the greatest number of securing the universal human goals. The Scots and the Moderns both provide analyses of successful spontaneous orders that rely on a principle of epistemological efficiency. Stability of expectations and spontaneous order are intimately linked to questions of knowledge: they add to our knowledge of the world by reducing uncertainty. At the same time the order provided by the core social spontaneous orders – law, science, morality and the market – has allowed the spontaneous development of a series of institutions – the division of labour and knowledge – that allow a growth in the cumulative sum of human knowledge through specialization. The growth of experience operates chiefly through the efficient adjustment of individual specialists to their particular circumstances and to each other. This efficient adjustment allows a benign equilibrium to form that meets the universal desire for secure subsistence. It is dependent on each individual being able to assess their position and to act accordingly to further their interests. This line of thought, so the spontaneous order theorists believe, provides an 'obvious' or instrumental justification for a liberal market economy. If the greatest number possible are to enjoy the satisfaction of universal human goals, then the market is the most efficient means that we have yet discovered to secure this. The critique of socialist planning and government intervention in the economy that the spontaneous order theorists provide is not grounded in the idea that they ought to pursue fundamentally different values. Rather, drawing on their spontaneous order analysis of social order, they assert that intervention and planning can never act as efficiently as freedom to secure goals that are universally held. If efficient adjustment and order are universally desirable, then freedom within a liberal institutional and legal framework represents the most epistemologically efficient means of attaining the key human goal of subsistence and material comfort. The point is not that other political approaches are 'immoral' or mistaken in their values, but that they fail to understand the nature of the social world and the implications of this for humanity's ability to shape its environment.

The spontaneous order theorists believe that success in securing the universal human goals can be indicated by increased levels of population and by rising living standards. This growth is dependent on the efficient use of resources and the co-ordination of the use of human knowledge in an efficient manner. Efficient mutual adjustment in the stable context of a system of general rules allows this specialization: freedom under the law and market exchange being the most efficient means yet discovered to secure social progress. Moreover, because individuals are capable of learning through a trial and error process, and of imitating the behaviour of those whom they perceive to be successful – and because they seek the same universal goals – then there will be a tendency, given the choice, for individuals to adopt those cultural practices which best secure their aims. While Hayek's theory of group selection appears crude and unfinished it represents an attempt to account for the process of cultural evolution in terms of spontaneous order and the growth of human knowledge. It might even be that the next generation of spontaneous order theorists will extend the model of cultural imitation and order-seeking to provide a conjectural history or composite model that accounts for the process of social change associated with globalization.

Indeed Virginia Postrel, in her *The Future and Its Enemies* (Postrel 1998), has begun to do just that. By examining globalization through the spontaneous order approach, Postrel argues that the key divide in modern political thinking is not between 'left' and 'right', but between those who accept a 'dynamist', spontaneous order vision of globalization and those who hold a 'stasist' view that opposes the instrumental justification of liberalism and the market. This analysis views 'stasists' – whether Conservative, Green, Socialist or Nationalist – as engaged in a campaign to prevent the global spread of the institutions and practices that we have referred to as representing the invisible hand.

This study has attempted to identify the constituent elements of the spontaneous order approach to social theory and to examine how the application of the approach can be coupled with an invisible hand argument to produce an instrumental justification of liberal values grounded in an assertion of their efficiency in securing a series of key, universal human goals. Its novelty has been in analysing spontaneous order as a distinct approach to the social theory of science, morality, law and government, rather than as the offshoot of a particular economic theory. Indeed the conclusion drawn is that the invisible hand argument that provides an instrumental justification of freedom depends on all of these institutions. The invisible hand is a series of evolved social institutions that allow the efficient discovery and co-ordination of knowledge in the pursuit of the human desire for material comfort.

Evolved institutions, such as morality, property and law, form the framework that allows the generation of benign spontaneous orders. And the mechanism of the invisible hand is that which creates benign spontaneous orders as the result of the co-ordination of human activity and the harmonization of the unintended consequences of human action. The generation of conventional general rules of morality and law; the common valuation of money and its use to provide price signals to co-ordinate economic effort; the pursuit and method of science; and indeed the human mind itself are all spontaneous orders that form a part of the invisible hand argument in favour of liberal principles. The spontaneous order approach, and the invisible hand arguments that draw upon it, constitute a definite approach to social theory that typifies a particular branch of classical liberalism. It is this approach that is Adam Smith's greatest legacy to political philosophy.

Notes

1 Spontaneous order in liberal political thought

1 Leading Hayek to refer to himself as an 'old Whig' (Hayek 1960: 409). Sufrin (1961: 202) believes that 'British Whiggery' is Hayek's ideal.

2 Gissurarson lists the key aspects of conservative liberalism as 'spontaneous order', 'anti-pragmatism', 'traditionalism', 'evolutionism' and 'universalism' (Gissurarson 1987: 11–13). Legutko (1997: 162), however, rejects the label of liberal conservative or conservative liberal as oxymoronic: a view shared by Barry (1979: 197) who points out that modern conservatism is descended from the Whig thought of Burke, in which case spontaneous order theorists are not mixing liberal and conservative thought but continuing the development of the Whig tradition.

3 Norman Barry (1982: 12) places the beginning of Western ideas of spontaneous order in the work of the 'School of Salamanca', or what Hayek refers to as the 'Spanish Schoolmen' (LLL vol. 1: 170 n. 8–9). The Schoolmen, writing between 1300 and 1600, are credited with the creation of a subjective theory of value that they applied to economics and in particular to money. In the writing of the Spanish Schoolmen there appeared the first conception of the idea of a self-regulating market based on the subjective valuation of goods (LLL vol. 1: 21). For a discussion of the Schoolmen and their analysis of 'automatic equilibrating processes' see Hollander (1973: 27). This view would later be taken up by the Austrian School of economists of which Hayek was a member.

  Matthew Hale's support of the English Common Law system is referred to by Hayek as having influenced Mandeville and Hume (LLL vol. 1: 22), but Hayek fails to provide sufficient evidence of the relationship, or indeed of Hale having developed a significant spontaneous order approach, for him to be considered as a serious candidate for inclusion in the tradition.

4 Burke (1985: 60–1) also rejects any direct relationship of influence between Mandeville and Vico, arguing that, as Mandeville wrote in English (a language with which Vico was unfamiliar), it is likely that his work escaped Vico's attention.

5 In recent years a debate has raged over Mandeville's position in the history of economic and political thought. Some, following Jacob Viner (1958), have viewed Mandeville as a mercantilist while others, following his editor Kaye (1988) and Rosenberg (1963) have instead viewed him as a precursor of laissez-faire thought. This debate touches on our concerns as Viner, Horne (1978) *et al.* have argued that far from advocating a spontaneous order approach to social change Mandeville instead argued for the intentional intervention, or 'dextrous management' (Dickinson 1975: 93), of skilful politicians, to ensure socially

beneficial outcomes. Under this view benign social consequences are not the result of an unintended consequence model of social change, but are instead the product of the deliberate channelling of individual action brought about by the intervention of politicians. This Viner-inspired argument is consonant with reading Mandeville as a paternalistic mercantilist: the view being that self-interest leads to beneficial results only when individuals have their actions constrained and guided by the intentional manipulation of politicians. In contrast to this view, Goldsmith has developed Kaye's reading of Mandeville and has argued that the passages which support the Viner interpretation can, in fact, be read to support a reading of Mandeville which views him as firmly within the spontaneous order tradition. Under this interpretation Mandeville's skilful politicians are not literal figures, but rather are 'an elliptical way of pointing to a gradual development whose stages we may not know but which we can reconstruct conjecturally' (Goldsmith 1985: 62). By viewing Mandeville in this light, as a conjectural historian, we bring him closer to the Scottish Enlightenment in terms of approach. There appears to be a widespread acceptance that Mandeville's thought influenced the Scots in a significant manner (Horne 1978: 33, 71, 92, 98; Goldsmith 1985: 101; 1988; Hundert 1994: 58, 83, 220–1, 219–36; Hayek 1984: 176–94), there is also considerable agreement that Mandeville was one of the first to deploy an unintended consequences model of social understanding (Hayek 1984; Hundert 1994: 77–8, 249; Goldsmith 1985: 40, 62; 1988; Rosenberg 1963), and this alone is cause to place him within the tradition of spontaneous order.

 6 Norman Barry (1982: 20) also refers to Josiah Tucker (1712–99) as an exponent of spontaneous order, but the absence of a fully developed social or political theory utilizing ideas of spontaneous order allows us to exclude him from our definition of the tradition.

 7 Barry (1982: 29) includes in his exposition of the tradition two French thinkers, Frederic Bastiat (1801–50) and Gustave de Molinari (1819–1912), and, although both advance arguments in favour of laissez-faire principles, neither appears to advance a gradualist spontaneous order style approach. The absence of an evolutionist, or gradualist, element in the thought of either of these thinkers, together with their rationalistic outlooks, makes their justification of freedom categorically different from that of the tradition of spontaneous order.

 8 For discussions of the Scots' influence on the Mills see Hayek (1960: 61), Forbes (1954: 664–70) and MacFie (1967: 19, 141, 145). MacFie (1990: 12) also argues that the Scots' approach to social matters is fundamentally at odds with that of Benthamite utilitarianism.

 9 See Hayek (LLL vol. 1: 152 n. 33); (LLL vol. 3: 154); (1960: 59, 433–4 n. 22).

10 It is worth noting here that Menger criticizes Adam Smith in various places for not being a spontaneous order thinker and for excessive rationalism (Menger 1996: 153–203; Rothschild 2001: 65). However, the critique is based on Menger's own, particularly idiosyncratic, reading of Smith and is rejected by Hayek as mistaken.

11 For historical background on the revival see Green (1987) and Graham and Clarke (1986).

12 Such approaches are rendered somewhat redundant when a thinker such as Hayek openly admits: 'But what I told my students was essentially what I had learnt from those writers and not what they chiefly thought, which may have been something quite different' (Hayek 1978: 52 n. 2).

13 Polanyi writes to Hayek: 'I have often disagreed with your views, but have done so as a member of the same family, and I have always admired, unfailingly, the power of your scholarship and the vigour of your pen'. (Letter from Polanyi to

Hayek dated 2/2/68, Friedrich A. von Hayek papers, Box no. 43, Hoover Institution Archives.)

14 Most studies of the idea of spontaneous order begin with an examination of the paradigmatic example of the market. However, because we are interested in the broader approach to the social sciences we shall proceed from the conception of science involved, following it through the explanation of the development of social institutions (including the market). With this aim in view the analysis will dwell primarily on primary texts from both periods. Occasionally we will engage with specific critical literature in the body of the text, but for the most part discussion of secondary literature will take place in footnotes.

15 The kind of argument suggested by Gissurarson where spontaneous order 'might not tell us what is desirable, but it may tell us what is definitely undesirable' (Gissurarson 1987: 65); and Gray's assertion that the idea of spontaneous order might not have a liberal content but may suggest liberal implications (Gray 1986: 124). Kley refers to Hayek's 'instrumental liberalism' (Kley 1994: 1) as a distinguishing characteristic of his approach: Hayek's argument is that liberalism and socialism are different methodologies with the same ends in view and can thus be compared in their relative success in securing those ends – a view shared by Mises (1978: 7–8). Walker (1986: 63) and Buchanan (1977: 31) both accuse Hayek of confusing positive and normative, descriptive and prescriptive arguments. This view, however, neglects the fact that Hayek does not pretend to produce a moral argument (Barry 1979: 5), nor does he seek to derive values from social science: rather he seeks to examine the pursuit of values by different methods – Socialism and Liberalism – (Barry 1979: 198–9). He presents a 'factual' argument for freedom (Connin 1990: 301).

16 Merton's analysis of *The Unanticipated Consequences of Social Action* restricts its attention to 'isolated purposive acts' (Merton 1976: 146) in terms of sociological analysis, though he highlights the fact that most social commentators have acknowledged the existence of the phenomena (Merton 1976: 145). The article itself stresses that there can be both malign and benign unanticipated consequences of action and that the reason for such consequences being unanticipated can be broken down into four categories: Ignorance, the 'Imperious immediacy of interest' and the roles played by basic values and self-defeating predictions in altering the circumstances in which the actions are undertaken.

His reference to the concept of unintended consequences with which we are concerned is made when he considers Adam Smith's invisible hand as an example of his second category of unanticipated consequences. His approach has been criticized by Karlson who argues that in referring to the 'unanticipated' consequences of action his analysis becomes unnecessarily vague. Karlson (2002: 27) instead argues that intention (or rather unintention) and not anticipation is the significant factor.

17 The sociologist Raymond Boudon, in his *The Unintended Consequences of Social Action*, broadens Merton's basic assumptions and examines what he terms 'perverse effects' (Boudon 1982: 8) in terms of social interaction. Most of Boudon's book is concerned with the sociological study of French education in the light of perverse effects, or unintended consequences; with his term perverse being applied to both benign and malign effects.

18 For comparisons of the invisible hand to the cunning of reason see Acton (1972), Ullman-Margalit (1978) and J.B. Davis (1983). Schneider (1967: xlvii) has argued that unintended consequence arguments were precursors of functionalism.

19 Polanyi had used similar terms applied to the same ideas in earlier works, especially the 1941 essay *The Growth of Thought in Society*, where he uses the terms

'spontaneous ordering', 'spontaneously arising order', and 'spontaneously attained order' (Jacobs 1998: 15).

20 See my forthcoming article in *The Elgar Companion to Hayekian Economics* for more on this point.

21 This stands against Gray (1986: 33–4) and Kley's (1994: 120) broader typologies which conflate unintended consequences, spontaneous order, evolution and the invisible hand.

22 There is a large critical literature that accuses Hayek of conflating what he referred to as the 'twin' ideas of spontaneous order and evolution. For example Barry (1982: 11), Gissurarson (1987: 61), Kley (1994: 38–9) and Petsoulas (2001: 17). The main charge here is that the mechanism for the emergence of a spontaneous order from mutual adjustment is not the same as the mechanism for the endurance of that order in an evolutionary, survival of the fittest process. While this criticism highlights one of the conceptual difficulties with Hayek's theory of cultural evolution (see Chapter 7), it does not have a direct bearing on the present study's use of the term evolution in a purely descriptive manner.

23 A view shared by Minogue (1985: 22) and Brown (1988: 135).

24 This understanding is the same as Rothschild's notion of the 'modern', as opposed to Smithian, invisible hand. She argues that the modern conception is characterized by three conditions: 'the unintended consequences of actions', 'the orderliness of the ensuing events', and 'the beneficence of the unintended order' (Rothschild 2001: 138).

25 The view that the invisible hand creates socially beneficial outcomes is also expressed by Petsoulas (2001: 34), Vaughn (1987: 997) and Elster (1989a: 96).

26 Nozick (1974: 20) refused to provide a detailed explanation of the concept, relying instead on a list of examples of its use by other writers.

27 Buchanan (1977) notes the possibility of 'spontaneous disorder' emerging from unintended consequences and appears to apply the term to two distinct notions: first, that unintended consequences produce no order and, second, that an existing order is disrupted by unintended consequences. Karlson (2002: 8, 23) shares a similar view.

28 It is important to note here that invisible hand arguments are not a form of naive identification of interests argument. As Hayek notes: 'neither Smith nor any other reputable author I know has ever maintained that there existed some original harmony of interests irrespective of those grown institutions' (Hayek 1967: 100). The invisible hand is to be found in those social practices and institutions that have evolved by a process of unintended consequences in such a manner as to facilitate beneficial social outcomes. Social science, correctly undertaken with a spontaneous order approach, allows us to examine the nature of the invisible hand and those practices and institutions upon which it depends.

29 When Lessnoff writes: 'by this stage it is clear that Hayek's social theory is no longer a neutral account of the evolution of human social structures, but is a defence of a particular kind of evolved social structure, one incorporating private property and a market economy' (Lessnoff 1999: 155), he is acknowledging the shift from a spontaneous order explanation to an invisible hand argument.

2 The science of man

1 For the influence of Newton on the wider Enlightenment see Gay (1969: 126–87). Discussions of his particular influence on Smith and Hume can be found in Redman (1993), Hetherington (1983) and Wightman (1975).

2 See my article in the forthcoming volume *New Voices on Adam Smith* for more on Smith's theory of science.

3 Both Smith and Hume are clear that this process, the desire to explain wondrous events in terms of science, is one which arises only after some economic progress has occurred. In simple societies this sense of wonder often invokes a mystical or religious explanation, but when a society materially advances and frees itself from the immediacy of savagery to such a degree as to support intellectual enquiry the reliance on miracles as explanatory devices gives way to rational enquiry.

4 See 'When the sentiment of the speaker is expressed in a neat, clear, plain and clever manner, and the passion or affection he is possessed of and intends, *by sympathy*, to communicate to his hearer, is plainly and cleverly hit off, then and then only the expression has all the force and beauty that language can give it' (LRBL: 25).

5 This process, for Smith, involved saving the best from previous systems and building on the work of the past (EPS: 53): Newton's 'standing on the shoulders of giants'. The importance of this observation will become clear as we proceed.

6 This overwhelming concern with causation indicates a significant feature of the Scots' (and spontaneous order thinkers more generally) work in that they are primarily engaged in a project of explanation rather than one of justification. Campbell and Ross (1981: 73) argue that: 'Smith's works are primarily analytic and explanatory'; while Gee (1968: 286) refers to Smith's 'definitional approach' and Becker (1961: 13) believes that his aim is to 'explain' society. Rendall notes that what distinguishes Ferguson from Smith and Hume is his greater concern with justification rather than analytical explanation (Rendall 1978: 149).

7 For a discussion of this see Rendall (1978: 123).

8 This is a point that will become particularly significant later in relation to Smith's analysis of sympathy.

9 Hume also draws on notions of comparison in his theory of aesthetics. In the essay *Of the Standard of Taste* (EMPL: 238, 243) he links experience to comparison in the role of the critic. Here he is keen to argue that it is possible to develop a universal standard of taste when arguments are made in relation to comparison and evidence. This provides an objective standard of taste. Where rational comparison is absent we are left solely with subjective value judgements with which it is impossible to argue. There is here some considerable similarity with the methodology of conjectural history: the objective standard (equilibrium) arising from rational comparison and leading to the generation of generalized principles.

10 See Broadie (2001: 67–8).

11 The Scots also highlight the need for detachment in the conjectural historian's attitude to other cultures. They were aware of the difficulties of approaching other cultures through the preconceptions of one's own and cautioned against the danger of ethnocentric bias. See Ferguson (1994: 184; 1973 vol. 2: 142) and Dunbar (1995: 152, 164).

12 As Skinner (1965: 5) would have it: 'the peculiar nature of their history lies in the link which it establishes between the constant principles of human nature and the changing environment of man'.

13 What Berry (1982: 59, 61) refers to as 'underlying functional universalities', or in Hume's terms 'constant springs' that allow the development of a science of man.

14 Lopreato (1984: 53) criticizes this approach in that it posits an unsubstantiated unilinear view of progress. He argues that the idea that Native Americans are at the same level as our ancestors is inaccurate because they lack the characteristics which turned our ancestors into us. This view, however, misses the point, in that the nature of conjectural history is what Hayek would call a composite

model and not an assertion of historical fact. It does not depend on a unilinear model of progress (Höpfl 1978: 24) because: 'the subject of conjectural history is not this or that society, or [still less] the human race, but the *typical* "society", "nation" or "people"' (Höpfl 1978: 25, his italics). Conjectural history is a process of classification, or ordering, used to make sense of the world and not to assert necessary or deterministic notions of development (Höpfl 1978: 39).

15 See Smith (TMS: 156) where he argues the futility of appeals to God in politics. Haakonssen (1990: 205) and Raphael (1979) both note Smith's preference for secular modes of argument. We see here the beginning of a line of argument that is characteristic of the spontaneous order approach and which distinguishes them within liberalism. They are wary of overtly justificatory arguments, and instead base their claims on explanatory theories.

16 See Ferguson (ECS: 8, 9, 11) where he repeatedly uses phrases such as 'no record remains', 'we must look for our answer in the history of mankind', and 'some imaginary state of nature'.

17 See also Dunbar (1995: 17). In a similar argument which enlarges on Ferguson's point, Dunbar argues that man is social before he is rational (Dunbar 1995: 16), indeed, that his intellectual powers are derived chiefly from the influence of socialization within the group.

18 We see here that the Scots clearly commit themselves to an explanatory, and not a justificatory project – a point that will become increasingly apparent in the following chapter.

19 See Millar (1990: 3), Hume (EMPL: 59) and Ferguson (1973 vol. 1: 168).

20 Smith highlights this by noting that in all polytheistic religions it is irregular events which are related to deities (EPS: 49), that is to say events which inspire wonder in the savage by exceeding the bounds of his habitual experience force him to have recourse to explanation in terms of divine intervention. See Chapter 1 where Macfie applies this to Smith's discussion of the invisible hand of Jupiter.

21 Though Hume (EMPL: 15) also rejects them.

22 Smith follows this line of thought when he contrasts Thucydides with Herodotus, arguing that the former is the superior historian because he does not digress on the personalities of the individuals that he writes about, but rather writes a fluid narrative (LRBL: 94). This does not, however, imply that the Scots adopt a form of methodological holism. Both Lehmann (1930: 157) and Vanberg (1986: 80) assert that the Scots are committed to a form of methodological individualism. Jones (1990: 6–7) notes that: 'Hume is loathe to consider groups of men as agents, in the strict sense: individual men are agents, motivated by their passions rather than by reason and reflection.'

23 See also Hume (EMPL: 481).

24 See Ferguson (ECS: 121), Dunbar (1995: 62), Hume (EMPL: 284) and Millar (1990: 5).

25 For similar arguments see Ferguson (ECS: 120) and Hume (EMPL: 125, 260).

3 The science of morals

1 Walton (1990: 38) notes the explanatory nature of the Scots' approach and suggests that Hume's thought provides a 'natural history of moral psychology'. Similarly, Haakonssen (1982: 205) highlights the psychological explanation that lies behind Smith's work viewing: 'his sympathetic theory of morals as a general framework for how a common morality emerges in a given society'.

2 See Hume (THN: 108) where the terms custom and cause and effect are used interchangeably: 'connected by custom, or if you will, by the relation of cause and effect'.

3 Non-deliberative in the sense of gestalt psychology. Which is to say that habits are not unthinkingly performed, but rather they represent 'subsidiary' knowledge drawn upon in the performance of a 'focal' activity. Instead of proceeding by a deliberative rational analysis of the situation, we draw on 'rules of thumb' developed from experience to assist us in attaining our aims.

4 See also Hume (THN: 654): 'We can give no reason for extending to the future our experience in the past; but are entirely determined by custom, when we conceive an effect to follow from its usual cause'.

5 Sociability, then, is a feature of the underlying universalities that make science possible. See also Swingewood's view that morality, for the Scots, was 'preeminently social' (Swingewood 1970: 169).

6 Lehmann (1930: 48) refers to this as the 'fact of society'; while Bryson (1968: 148) argues that, for the Scots, empiricism proves sociability.

7 This repetition is posited on the success of the action in the first place. That is it must effectually fulfil some purpose if it is to be repeated. It is this idea that underlies Hume's notion of utility.

8 See also Hume (EMPL: 50), where he argues that humans are attracted by novelty.

9 A custom develops as an inter-subjective equilibrium of behaviour that becomes settled as the group becomes habituated to it.

10 Sugden (1989: 87) offers a defence of Hume's views on convention by stressing the role that they play in stabilizing expectations. He argues that: 'The belief that one ought to follow a convention is the product of the same process of evolution as the convention itself' and relates this tendency to equilibrium to the natural propensity to seek order.

11 See Ferguson (ECS: 23); Smith (TMS: 84) and Hume (THN: 363).

12 Smith makes a similar point regarding such circles of concern and their relation to the strength of our feelings for others (TMS: 86, 142). Ferguson also dwells on the point (Ferguson 1994: 247; 1973 vol. 1: 30; 1973 vol. 2: 293).

13 James Otteson (2002: 183–9) provides a detailed analysis of the significance of what he calls the 'familiarity principle'. The analysis here is in fundamental agreement with his approach.

14 The full details of Smith's views on the implications of this are to be found at (TMS: 135–7). Hume (THN: 416) also makes a similar point about fingers.

15 It is from this point of view that Smith (TMS: 229) notes that the love of one's country does not arise from a generalized love of humanity. Rather it grows outward from our attachment to and concern for what is close to us. See also Hume (THN: 481).

16 In this sense they form a part of the invisible hand that produces socially beneficial results from the interaction of individuals in pursuit of their own purposes.

17 The wise man, Hume argues, 'will endeavour to place his happiness on such objects chiefly as depend upon himself' (EMPL: 5).

18 What Smith is advancing here is an argument which parallels, as we will see below, his epistemological argument in relation to economics (and thus represents an invisible hand argument). An individual has the most accurate access to knowledge of his particular circumstances: he is most intimately familiar with both his situation and that of those related to him. As a result, so long as he does not actively put down others' attempts to achieve their goals, he is the person best placed to provide the most efficient outcome for himself and his intimates. Individuals then are best fitted to adapt to their circumstances. See Smith (TMS: 83, 138).

19 Wilson (1997: 18, 45) links this natural sympathy to our innate sociability.

20 Broadie (2001: 104) and Skinner (1996: 60) both highlight the epistemic role of

the impartial spectator in reaction to our limited knowledge. While Mizuta develops the view that: 'The effort to be sympathized with by moderating the individual's own emotion should be increased as the society in which he lives becomes greater and as the distance between him and the spectators increases' (Mizuta 1975: 121).

21 Ferguson notes that the force of this aspect of human character is such that: 'Without any establishments to preserve their manners ... they derive, from instinctive feelings, a love of integrity and candour, and, from the very contagion of society itself, an esteem for what is honourable and praiseworthy' (ECS: 156). Ferguson's language here reveals that this argument is an invisible hand argument where a socially beneficial result is produced without purposive organization.

22 Christina Petsoulas presents an interesting analysis of this approach. She argues that: 'For Smith, imaginative sympathy is the mechanism whereby men with different experiences, occupying different positions, and frequently having conflicting interests, are able to develop common rules of conduct' (Petsoulas 2001: 152). These rules of conduct become 'crystallized common standards of moral evaluation' (Petsoulas 2001: 153) as a result of the repetition of mutual sympathetic approval. From here Petsoulas introduces a new line of argument, such that: 'men *purposefully* employ the psychological propensities of the imagination (custom or habit formation), first to *discover* rules and institutions, and subsequently to *enforce* them' (Petsoulas 2001: 109, her italics). This leads her to assert that: 'Though the means by which we arrive at impartial moral judgements is still sympathy, it is a form of sympathy mediated by *conscious reflection*' (Petsoulas 2001: 121, her italics). This argument does not appear, at first glance, to deny that sympathy, moderated by the impartial spectator, is a facet of human nature in Smith's theory. Instead Petsoulas suggests that men make conscious use of the impartial spectator in a deliberate attempt to provide common moral standards of evaluation. This view is clearly mistaken. It confuses the individual's conscious reflection on a particular moral issue (the conscious evaluation of a moral dilemma through an interaction of reason and sympathy) with a desire purposefully to create a system of enforceable moral standards. Though an appeal to the impartial spectator is obviously in some sense deliberate and involves conscious reflection, the purpose of this reflection is not the creation of a moral code. For example, the human desire to stabilize expectations is not the conscious motive behind our moral deliberations, rather it explains why we have moral deliberations. In this sense the moral code is an unintended consequence of a series of sympathetic reactions and conscious reflection on particular cases of sympathetic approval.

23 What Griswold (1996: 191–2) calls the 'contextuality of the moral sentiments'.

24 Kerkhof (1995: 221) has noted the function of shame as an inhibitory force in the Scots' moral theory.

25 See also Ferguson (1973 vol. 1: 58): 'As nature seems to try the ingenuity of man, in a variety of problems, and to provide that the species, in different countries, shall not find any two situations precisely alike; so the generations that succeed one another, in the same country, are, in the result of their own operations, or the operations of those that went before them, ever made to enter upon scenes continually varied. The inventions of one age prepare a new situation for the age that succeeds; and, as the scene is ever changing, the actors proceed to change their pursuits and their manners, and to adapt their inventions to the circumstances in which they are placed.'

26 Remember that the Scots' interest at this point is explanatory and not justificatory.

27 Ferguson (1973 vol. 2: 232) argues that convention 'may be supposed almost coeval with the intercourse of mankind'.
28 See also 'the object of prudence is to conform our actions to the general usage and custom' (THN: 599).
29 As with propriety, so prudence also forms a part of the invisible hand.
30 See James Otteson's *Adam Smith's Marketplace of Life* (2002) for a detailed study of this aspect of Smith's moral philosophy. A recent approach to civil society which adopts a similar view is to be found in the work of Edward Shils (1997).
31 See Smith (TMS: 209–11), Hume (EMPL: 398–9) and Ferguson (ECS: 135).
32 See also Smith (EPS: 37). Ferguson questions the utility of the practice, believing that far from limiting population it created a perverse incentive which removed all restraint on sexual activity leading to a gradual growth in population (ECS: 135): an example of a possibly malign unintended consequence of customary human behaviour.
33 See also Hope (1989: 86–7). For a similar analysis applied to the use of Latin in the Roman Catholic Church see Smith (WN: 765). Smith (TMS: 199) notes that customs can be viewed as absurd from the outside and that our socialization within particular custom can effect our judgement of other customs (TMS: 148); but he also believes that by examining the circumstances which produced a particular practice in the light of their relation to both sympathy and utility, we are able to undertake a process of immanent criticism and to form judgements as to their success and moral value. Petsoulas (2001: 115) develops a similar argument with reference to Hume. She argues that Hume's notion of conscious reflection on evidence provides an external standard by which to judge and alter evolved behaviour patterns. Thus the philosopher is able to engage in immanent criticism of a moral practice, but this criticism is not the model for the explanation of the change in moral beliefs. Once again purposive rationality is downplayed.

4 The science of jurisprudence

1 Smith (WN: 710–11) makes a similar point. Note here that the Scots did not believe that they were providing a moral justification of a particular form of property, rather they believed that they were providing an explanation of the historical development of the phenomenon (Bowles 1985).
2 Kames also developed a stadial theory while Ferguson's 'highly idiosyncratic' (Meek 1976: 154) analysis appears to operate with three, rather than four, stages: savage, barbarous and polished. See Kettler (1965: 228), Lehmann (1930: 81–6) and Hill (1997: 679). The origins of the 'four stages' approach have been traced to Grotius (Meek 1976: 14) and the Physiocrats (Meek 1971). However, Bowles (1985: 197) points out that the Scots' explanatory approach prompts us 'to ask historical questions rather than the moral questions of the natural law framework'. See also my contribution to the forthcoming volume *New Voices on Adam Smith* which complements the argument of this chapter.
3 Smith undertakes similar case studies, in particular devoting a chapter of the *Wealth of Nations* to a 'four stages' analysis of the development of the military that forms part of his argument for a standing army (WN: 689–708).
4 It should be noted that the 'four stages' is not a deterministic model of inevitable development, but rather represents an attempt at explanation through the medium of conjectural history (Broadie 2001: 76; Skinner 1996: 183; Harpham 1983: 768–9). It is also, as Cropsey notes, significant in its downplaying of politics in favour of an underlying economic understanding of the forces behind social change (Cropsey 1957: 57).

5 Heilbroner (1975: 527) correctly states that population growth is the force behind the change between stages, but it cannot, by itself, explain the development. In Meek's (1976: 213) terms hunger prompts the search for new knowledge. Like the mercantilists (Hollander 1973: 58–65), the Scots viewed population growth as an indicator of progress. Danford (1990: 183–6) has argued that Hume's essay *Of the Populousness of Ancient Nations* is a contribution to the debate over the superiority of classical models of freedom to modern 'commercial' freedom. Hume uses population levels here to suggest the superiority of the modern approach. See also Reisman (1976: 146). Similarly, Spengler (1983) argues that Smith regarded a decline in infant mortality as an indicator of economic advance.

6 The social change brought about by the change in the mode of subsistence is thus an unintended consequence of the development of new ways of procuring subsistence (Meek 1976: 224).

7 Hont (1987: 254) suggests that the fourth stage differs from those prior to it in that it does not refer to a productive process related directly to the attainment of subsistence. Rather, trade, which is present in all four stages, comes to represent the chief means of securing subsistence through interdependence. Meek (1976: 227) also notes that the change to the fourth stage differs to previous changes, in that it is the development of a factor that has always been present [trade] that is significant, rather than the acquisition of a practical skill of production.

8 This having been said the development of property proceeds according to the unintended consequences model. See Hume (THN: 529). In addition, Hume's use of the terms natural and artificial is qualified by the assertion that mankind is by its nature an inventive species (THN: 484).

9 Though Smith (TMS: 179, 188) questions Hume's over reliance on utility as an explanatory factor.

10 Hume stresses that, though a system of property is in the public interest this is not the motive which prompted its establishment. It is the self-interest of individuals who adjust their behaviour with their own gain in view. See 'This system, therefore, comprehending the interest of each individual, is of course advantageous to the public; tho' it be not intended for that purpose by the inventors' (THN: 529). Individuals do not purposively set out to create a system of property, rather it is an unintended consequence of the interaction of self-interest and the desire for stability of expectations.

11 And thus form a part of the invisible hand argument for the generation of benign spontaneous orders.

12 As Baumgarth (1978: 12) notes: 'Rules, whether legal or social, in the sense of mores, have as their task the reduction of uncertainty, at least avoidable uncertainty'.

13 There is an implicit question here as to what extent the Scots believed that their explanation of the origins of property and government served as a convincing justification of particular forms of government or property. See Bowles (1985).

14 'Such results are attained through the activities of man in the mass; results of which the individual is largely unconscious but which he may later recognize' (Skinner 1967: 43). This view has led some commentators to view Smith and Hume as system utilitarians (Campbell and Ross 1981: 73).

15 Smith (LJP: 211, 405) and Ferguson (ECS: 98) make similar points within their 'four stages' analysis. See also Rosenberg (1976), for a discussion of Smith's views on the role of judges and chiefs as a manifestation of the division of labour.

16 Hume (EMPL: 40) makes this point when he refers to all government being founded originally on force or usurpation. While Smith (LJP: 321, 402) argues

that submission to government is grounded on opinions of authority and of utility that exist in a complex interrelationship.

17 Carabelli and De Vecchi (2001: 241) view the role of the judge here to be that of a 'gap-plugger' in the same sense as we saw in Smith's views on the role of the philosopher or scientist. Stein (1996: 165) refers to the Scots' 'dynamic' conception of law as an adjustment to circumstances. Or, as Livingston would have it: 'law, like language or any other profound Human convention, evolves spontaneously, guided by custom and tradition. It is not due to the insights of speculative philosophers and the craft of constitution makers' (Livingston 1990: 129).

18 For the Scots on man's naturally progressive nature see: Dunbar (1995: 4), Ferguson (1973 vol. 1: 257; 1973 vol. 2: 85), Kames (1751: 97, 100; 1776: 64; 1774 vol. 1: 230), Stuart (1768: 217) and Smith (WN: 540).

19 We will deal later with the question as to the Scots' views on the possibility of decline as well as progression (ECS: 198, 204). Suffice it to note that the Scots viewed progress as favourable, but did not believe that it was inevitable or assured (Höpfl 1978: 37).

20 What Lehmann, referring to Ferguson, calls 'an ever-increasing fund of experience' (Lehmann 1930: 69).

21 Chitnis stresses the explanatory nature of the Scots' views here. He argues that their conception of progress implies 'no qualitative judgement' (Chitnis 1976: 96). They are providing a retrospective explanation of social change not a justification of a particular form of society.

22 See Ferguson (ECS: 7, 10, 13; 1973 vol. 1: 18–19).

23 Again unintended consequences come to the fore. We see that the progressive elements in society: law (Millar 1990: 164); contracts (LJP: 205); the division of labour (WN: 25); liberty (LJP: 271); and commerce and manners (WN: 412) – are spontaneous orders produced by the process of unintended consequences. They thus represent an invisible hand argument regarding the formation of benign orders.

24 Smith (TMS: 226) goes so far as to argue that its importance is such that it outweighs other moral concerns because it is essential to the very existence of human society.

25 Ferguson is similarly scathing as to the ability of a government to act successfully in economic matters arguing that statesmen can 'do little more than avoid doing mischief' (ECS: 138). However, as we will see later, Ferguson expresses doubts about Smith's restriction of the scope of government action.

26 This, as we will see in the following chapter, is an invisible hand argument.

27 One of the chief reasons the Scots doubt the effectiveness of such sweeping intervention is related to their unintended consequence approach. If progress is posited on the unintended consequences of social interaction producing an effective utilization of the cumulative sum of human knowledge, then no one man can hope to form a vision of a more efficient society from his own, limited, reason (EPS: 318–19; EMPL: 351).

28 Note how Hume (EMPL: 30) stresses that malign unintended consequences may be produced by well-intentioned actions, he cites the example of Brutus.

5 The science of political economy

1 For a condensed version of the arguments in this chapter, see my 'Adam Smith on Progress and Knowledge', forthcoming in the Routledge volume *New Voices on Adam Smith*.

2 A point also highlighted in Ferguson (ECS: 172) and Hume's (THN: 485) respective analyses of the division of labour.

3 The suggestions of utility underlie Smith's economic analysis of the division of labour in the *Wealth of Nations* and *Lectures on Jurisprudence* (LJP: 351), but he also provides a psychological account of specialization that can be related to those human tendencies that prompt humans to science. We have already noted that the Scots discern a natural human propensity to seek order in the understanding of the world. From this they drew a notion of the human mind as functioning by classification (EPS: 38–9), and as this classification naturally develops in line with experience, so the differentiation of experience that occurs creates different fields or objects for human study. However, the psychological explanation of the pursuit of specialist knowledge is linked to both utility and sympathy. Smith argues that humans naturally admire the knowledge of specialists (TMS: 20), and moreover they see how specialization has provided these people with a safe route to wealth and reputation (TMS: 213). There is a sense in which we pursue specialized knowledge from an emulation of the rich and successful (TMS: 55). Inspired by their success we seek to acquire knowledge and express our talents in order not only to secure financial reward, but also to enjoy the acclaim that goes along with expertise (TMS: 181).

4 Spengler (1983) and Brown (1988: 79–80) both make this point.

5 There is a link here to the changes between modes of production in the 'four stages' theory: an increase in population allows specialization and increases the likelihood of useful observations being made that, in turn, lead to a further increase in population.

6 Smith (WN: 20–1) links this theme to an unintended consequences argument about the motivations of workers: where workers improve a machine in order to reduce the amount of labour required of them.

7 Smith (TMS: 336) argues that this phenomenon, and the interdependency which it creates, is a further reason why we admire specialists.

8 Rosenberg (1965: 128–9) agrees with this view, and develops it into an argument that a decreasing intelligence in particular labourers, resulting from their concentration of attention on a particular task, need not prevent the continuation of overall technical progress. His view is that the division of labour represents a process of simplification in reaction to complexity, the result of which is that 'the collective intelligence of society grows as a result of the very process' that restricts the breadth of individual knowledge (Rosenberg 1965: 134–5).

9 Smith (WN: 22–3) goes on to highlight this by listing some of the chain of interconnections.

10 See also Hume (EMPL: 324). Note here that Smith's support for free trade is a result of conclusions drawn from his descriptive analysis of the operation of the division of labour. It is in this sense an instrumentalist and not a normative argument.

11 For Smith on reputation, trade and the advance of manners see (TMS: 57, 213; LJP: 13). See also Chapter 3 on prudence and propriety. Hirschmann examines the historical development of this style of argument, referring to its core assumption that views 'money-making as a calm passion' (Hirschmann 1977: 63).

12 The so-called Adam Smith problem: for a discussion of which see Dickey (1986), Teichgraeber (1981) and Otteson (2002). James Otteson's treatment is particularly revealing because, like the present study, he focuses on the model of the unintended generation of social order that is found throughout Smith's work.

13 See also Smith (LJP: 572). There is an implicit epistemological argument here that Vernon Smith picks up on. He argues: 'Not knowing of the invisible good

accomplished by the self-interest in markets, but knowing of the good we accomplish by doing things for friends, we are led to believe we can do good by interfering in the market' (Smith 1997: 30).

14 See Hume (EMPL: 280), Smith (TMS: 308–12) and Ferguson (ECS: 9).

15 For examples see Hume (EMPL: 84) and Ferguson (ECS: 53).

16 It is a feature of Smith's explanatory project that, though he may personally disapprove of excessive self-interest, he stakes no moral argument against it in his analysis of commerce. As Bishop notes: 'Smith did not feel the need to draw moral conclusions from the invisible hand argument because his view of natural liberty made it redundant' (Bishop 1995: 177).

17 This is highlighted by Smith's focus on the importance of 'self-command' (TMS: 25), and in Smith and Ferguson's cases is drawn from the classical tradition of stoicism.

18 See Smith's discussion of the difficulty in assigning motives to particular actions in the *Lectures on Rhetoric and Belles Lettres* (LRBL: 171).

19 Their analysis here is undertaken within the 'four stages' schema (Reisman 1976: 129–38).

20 See Millar (1990: 273) and Smith (WN: 389–90).

21 The inefficiency of slavery is discussed in Rosenberg (1965).

22 A similar point is made by Ferguson (ECS: 127) and Hume (EMPL: 278).

23 This is perhaps made most clear by the Scots' analysis of Magna Charta in terms of unintended consequences (Hume 1983 vol. 1: 437–44; Millar 1812 vol. 2: 80–1).

24 Smith (LJP: 71, 524) attacks the effectiveness and morality of both primogeniture and entailed legacies.

25 This, of course, is an example of an invisible hand argument.

26 A point that, we have noted before, was a vital step in the development of the division of labour.

27 See also Smith (WN: 366) and Ferguson (ECS: 124). Porta and Scazzieri stress that, for Smith, this interaction occurs within the context of a system of intersubjectively generated rules: 'Economic coordination presupposes a structure of beliefs, symbols and communicative codes, that is, a body [tradition] of mutual adjustments and interdependent decisions by which the outcome of social interaction is constrained' (Porta and Scazzieri 2001: 2).

28 Dunbar (1995: 77–8) refers to an 'undesigning hand' rather than an invisible hand, stressing that the order produced is a result of human action but not human design. Flew (1987) believes that Smith's invisible hand is an explanation of the co-ordination device which solves a knowledge problem within the economy.

29 For example, Evensky (1993) views the invisible hand as the hand of God; while Kleer (1995) sees God as the final cause behind all of Smith's work; and J.B. Davis (1983), Martin (1983) and MacFie (1971) all view the invisible hand as a metaphor for providence. Evidence for this view is generally drawn from Smith's occasional references to his belief in a Theistic conception of God (TMS: 105–6, 128, 166; 1987: 68), what Brown (1988: 136) has called his 'theistic platitudes'.

30 Others who reject the religious reading of the invisible hand include: Cropsey (1957: 27), who traces it purely to self-preservation; Camic (1983: 59–63), Haakonssen (1982: 205) and Rashid (1998: 219), who view the Scots as secularists with little interest in invoking God as an explanatory device; Rosenberg (1990a: 21), who claims that the invisible hand 'has nothing to do with divine guidance', but is instead the product of competition; and Flew (1985: 58), who argues that 'Smith's invisible hand is not a hand, any more than Darwin's natural selection is selection'.

31 Rothschild (2001: 117) admits that her evidence for these assertions is 'indirect'.

32 Rothschild (1994: 320–1) argues that by 1770 the idea of unintended consequences was a cliché used as a justification for policies against free trade: a
view that appears to neglect Smith's explanation of historical change and the
fact that his project was explanatory rather than justificatory.

33 Menger (1996: 131–5) offers a later, more detailed conjectural history of money.
It should also be noted at this point that the Scots, like Hayek, did not operate
with an idealized view of the market. Their analysis is concerned with 'real'
markets rather than with the construction of models of perfect competition.

34 Yet another example of the tendency which we have identified for humans to
seek simplification in reaction to complexity, an unintended consequence of
which is the development of still greater complexity.

35 Governor Pownall, in a letter to Smith, criticizes this distinction arguing instead
that there is only one real sense of price, market price (Smith 1987: 337–76). For
discussions of Smith's views on value see Hollander (1975: 315–16) and Vickers
(1975). Hutchison (1990b: 92–3) argues that, apart from the labour theory of
value, Smith's approach is largely subjectivist: while Paul (1977: 294) follows
Pownall's critique of the labour notion of value in the light of Smith's views on
trucking. Skinner (1996: 146–50) believes that the notion of labour value was
part of a 'vain' search for an absolute measure of value. This vain search only
really leaves the tradition of spontaneous order with the innovations of the Austrian economists. For a direct comparison of Hayek and Smith on the spontaneous emergence of equilibrium prices see Recktenwald (1990: 114).

36 Smith (TMS: 181) is ambivalent in his attitude to the force which moves the
market, he sees its value but notes also its possibly distasteful results. Once
again this underlines the explanatory nature of his spontaneous order approach.

37 In this sense the co-ordination achieved by the price mechanism can be understood as an expression of the invisible hand (Skinner 1990: 137).

38 See Smith (EPS: 311; WN: 687) and Hume (EMPL: 52).

39 Smith (WN: 522–3) qualifies this by noting that comparative advantage ought
not to be extended to those industries vital for the defence of the nation.

40 In Dunbar's terms the 'circle' of wealth 'widens' (Dunbar 1995: 365).

41 The Scots' views on luxury are part of a wider debate: suffice it to say here that
the Scots believed that luxury was not necessarily a debauching phenomenon
which destroyed virtue (EMPL: 276). Luxury was a relative concept (ECS: 232),
which could produce both beneficial incentives and possibly dangerous degrees
of avarice if made the sole focus of action (ECS: 109). See Berry (1994: 163–4).

42 For a discussion of the historical background to Smith's critique of mercantilism
see Coats (1975).

43 Smith's language here is particularly revealing about the nature of the misdirec-
tion caused by price perversion. Note the phrases 'taught to run', 'artificially
swelled' and 'forced to circulate' (WN: 604–5).

44 Competition prevents the self-interested businessman from exploiting his position to pervert the price mechanism. See Rosenberg (1990a: 21) and Teichgraeber (1986: 135).

45 Although Smith is on the whole dismissive of the alleged benefits of monopolies
he does see the case for some temporary monopolies being granted to those who
open up new areas of trade (WN: 754–5). However, he is clear that, like a copy-
right, such an indulgence should only be for a limited period after which free
competition ought to be allowed. Once again though Smith notes that the
granting of such privileges depends on the support and acquiesence of governments.

46 West (1975) and Heilbroner (1975) have produced the most notable discussions of the 'paradox' that can be identified in the Scots' attitude to the division of labour. Lisa Hill (1997: 683) identifies the 'paradox' in Ferguson's work as that between his faith in spontaneous evolution and the material progress afforded by commerce on the one hand, and his critique of the political institutions of modernity on the other. This paradox leads Kettler (1977: 439) to assert that: 'Ferguson cannot be simply classed with civic humanist pessimists or with historicist progressivists'. A view shared by McDowell (1983) who reads Ferguson's work as an attempt to create a fusion that balances republicanism and commerce into a 'commercial republicanism'.

47 Skinner (1996: 205–6) rightly notes that if, for the Scots, morality is the product of socialization and contact with others, then the division of labour has the potential to reduce the interaction on which the generation of conventional morality depends. It is as a result of this that the Scots stress the importance of socializing outside the workplace.

48 See Smith (WN: 723, 759, 780–1, 815; TMS: 222).

49 See Sher (1989) and Robertson (1985).

50 See Ferguson (ECS: 140–6).

51 See Ferguson (ECS: 155, 176), Hume (EMPL: 253) and Smith (LJP: 182, 226–7).

52 See Ferguson (ECS: 219), (1830: 468) and (1756).

53 See also Millar (1990: 236).

54 See also Ferguson (ECS: 175, 254–5) where he attacks the professional Chinese bureaucracy. Gellner (1994: 75) notes that Ferguson has far greater concerns about this political division of labour than he does about the possible ill-effects on industrial workers. This reading downplays the claims that Ferguson prefigures Marxian views on the alienation of industrial workers (Brewer 1986).

55 MacRae (1969: 23) points out that, though Ferguson highlights the possible negative results of the division of labour, he does so while maintaining the spontaneous order approach. The negative results take the form of malign unintended consequences of the development of commercial society.

56 Goldsmith (1988: 591; 1994) argues that the civic republican ideal was rendered obsolete by the advance of the division of labour, and that, by Ferguson's time, it was utterly impracticable as a guide to reform.

6 The evolution of science

1 Hayek admits his debt to Smith on this point in a note (Hayek 1967: 22 n. 1).

2 As Popper puts it science 'systematises the pre-scientific method of learning from our mistakes' (Popper 1994: 100).

3 It is precisely this point which leads Hayek to talk of the mind as operating with 'higher order rules' beyond our conscious perception. Polanyi shares a similar view, his concept of faith or belief as the basis of science is grounded in his assertion of 'the ubiquitous controlling position of unformalizable mental skills' (Polanyi 1969: 105–6).

4 Hayek (1979: 86, 158), Popper and Polanyi (1951: 77, 89) also make the point that there is no way that mankind can explain its knowledge. For it, logically, would have to 'know' more than its knowledge in order to explain it (an infinite regression problem). The same principle that leads Hayek to talk of the higher order rules that guide consciousness, but can never themselves be perceived. As a result the advance of science cannot be planned (Hayek 1960: 33), because it would mean having knowledge about that which we have yet to acquire knowledge of. A further corollary of this is that scientific method itself cannot be

proved in the strict sense but only upheld because of its continued success in stabilizing expectations (Hayek 1984: 256).

5 A point closely related to evolutionary psychology. See Cosmides and Tooby (1994: 328–9).

6 For a discussion of the implications of this for the methodology of social science see Lessnoff (1974) and Winch (1990).

7 Popper (1989: 133) refers to this approach as a 'holistic' error of 'naive collectivism' and argues, along with Hayek, that it must be replaced by a methodology that focuses its attention on the central role of individuals.

8 Compare with Winch (1990: 107).

9 For an analysis of Hayek's subjectivism and his relationship to methodological individualism see Caldwell (1994, 2004).

10 Having said this it is possible to be a methodological individualist without holding concerns about dispersed knowledge and unintended consequences: equally it is possible, though more difficult, to be concerned with unintended consequences and dispersed knowledge without being a methodological individualist.

11 Where Hayek and Popper diverge though is over the relationship of a social science thus constituted to the methodology of the natural sciences. Hayek argues that the social sciences differ from the natural sciences in their handling of their subject matter. The social sciences proceed, according to Hayek, by a subjective or 'compositive' method (Hayek 1979: 67). They build conceptual models in order to account for regularities in their complex subject matter rather than search for explicit universal 'laws' (Hayek 1967: 42). The conclusions of such a social science are necessarily subjective and limited in their application. While Popper agrees with much of this approach he also denies that it signals a difference in methodology between the social and natural sciences. Popper believes that this traditionally perceived difference between the methodology of the natural and social sciences is a mistake. He admits Hume's point about the difficulty of conducting experiments in social science, but he does not believe that this leads to a fundamental difference in methodology (Popper 1961: 9, 85). Popper argues that both natural and social sciences share the same underlying methodological approach in that they are both concerned with problem solving and proceed by a process of conjecture and refutation (Popper 1972: 185). He argues that this common commitment to a hypothetico-deductive approach indicates a 'unity' of method common to all science (Popper 1961: 130–1) and that the difference which leads Hayek to refer to the 'scientism' of applying natural science methods in the social sphere, is in fact based on a misunderstanding of the scientific method of the natural sciences by those thinkers (Popper 1966 vol. 1: 286). Once the inductive process has been rejected in all science, so Popper argues, there is essentially no difference in approach. That is to say that Hayek's compositive method for social science is at base, for Popper, compatible with his own hypothetico-deductive method of understanding of the nature of science (Popper 1961: 141). It has been widely noted that Hayek eventually became persuaded by Popper's argument over this and modified his views (Butler 1983: 145; Kley 1994: 44; Popper 1994: 140). Gray (1986: 12), however, doubts that Hayek was ever fully won over to Popper's view.

12 For discussions of Hayek as a conjectural historian see Walker (1986: 94) and Butler (1987: 124), both of whom suggest that the approach lies at the heart of Hayek's social theory.

13 A view endorsed by Hayek (1967: 4; 1984: 274, 325).

14 There is an obvious parallel here with Hume's views on the *Standard of Taste*, where critical discussion allows an objective standard to arise from subjective opinion.

15 Note here also the similarity to Smith's notion of the interpersonal generation of moral value through sympathy.

16 What is objective is that which has been inter-subjectively tested (Popper 1959: 44) leading to an instrumental justification of free debate, or a 'liberal epistemology' (Barry 1986: 115) along the lines of Mill's *On Liberty* (Gray 1989: 24).

17 In Oakeshottian terms they participate in the 'conversation' of science (Oakeshott 1991: 490) by pursuing 'intimations' from the tradition of scientific debate (Oakeshott 1990: 240).

18 The problem with such discipline is that it necessarily restricts the debate in the sense that, when discussing a current notion or theory, the debate will ignore or fail to pay attention to a contribution which is off subject (Polanyi 1969: 79). Polanyi discovered the truth of this assertion himself when his theory of adsorption was ignored for almost 50 years because it went against the established debate – even though it eventually became the established view (Polanyi 1969: 94). Dissent from the orthodoxy can be futile until such time as the debate shifts to a 'gap' in the consensus that the dissenter is able to exploit to surpass and explain it. Popper endorses a similar view when he argues that 'a limited amount of dogmatism is necessary for progress' (Popper 1994: 16).

19 Though as we have seen Popper doubts the value or significance of this division.

20 Though Hayek (1979: 19) exempts the Scots from the charge of scientism as a result of their rejection of constructivist rationalism.

21 All of which bears obvious similarities to Oakeshott's (1991) critique of rationalism.

22 Hayek's critique of socialism is not based on a dispute over values, but rather on an assertion that socialists have fundamentally misunderstood the nature of society in such a way that their proposed political and economic reforms are doomed to fail to achieve their stated aims (Gissurarson 1987: 66, 77).

23 This belief in the predictive powers of historical laws differentiates historicism from the conjectural history approach of the liberals. For the liberals conjectural history explains the past but does not allow the formulation of detailed predictions as to the future course of events.

24 Both Merton and Karlson note this, leading Merton to prefer the term the 'unanticipated consequences' of action rather than the unintended consequences (Merton 1976). Karlson rejects this terminology arguing that it is intention, and not anticipation that is the significant factor (Karlson 2002: 27). For example, we might anticipate that the free market produces collective benefits, but we do not intend the creation of these benefits when we act in market transactions.

25 Popper (1966 vol. 2: 96) notes that the likelihood of such a phenomenon is rare, so rare in fact that it might be an object of curiosity as to why there were no unintended consequences.

26 For a discussion of Hayek's use of the crystal example see my forthcoming article in *The Elgar Companion to Hayekian Economics*.

27 Popper (1961: 80) links this to 'holistic' approaches to the study of society. One cannot study the 'whole' of society because one's study would be a part of that whole leading to a problem of infinite regression. Note also how this line of thought compares to Hayek's views on the higher order rules that govern consciousness (Barry 1979: 14).

28 Popper (1989: 135) refers to reason as a 'tradition'.

29 It is because of this that Hayek argues that reason always works in combination with the non-rational (Hayek 1993 vol. 1: 32).

30 This is why Hayek constantly cites Hume's view that morality is not the product of reason (Hayek 1960: 63, 436 n. 37; 1967: 87; 1988: 8, 66).

31 This sense of utility is Hume's notion, not the later classical notion of Bentham and Mill which, as we noted above, Hayek rejects as a species of constructivist rationalism (Hayek 1993 vol. 2: 17). Gray (1986: 59) and Butler (1983: 16) refer to it as 'indirect utilitarianism' or 'evolutionary-system utilitarianism' (Gray 1989: 92).

32 This line of thought is also to be found in Oakeshott's analysis of the development of practices. A practice for Oakeshott, like a skill for Polanyi, is not consciously applied but rather emerges from use and can only come to be understood through examination of the use made of it (Oakeshott 1990: 120–2). As Oakeshott would have it: 'More commonly however, a practice is not the outcome of a performance. It emerges as a continuously invented and always unfinished by-product of performances related to the achievement of imagined and wished-for satisfactions other than that of having a procedure, and it becomes recognizable when it has acquired a certain degree of definition and authority or acknowledged utility' (Oakeshott 1990: 56).

33 Hayek even suggests that the term 'institution' possesses unfortunate suggestions of deliberate design and instead suggests the term 'formations' (Hayek 1979: 146–7).

7 The evolution of morality

1 As Shearmur notes, this approach views rules as constitutive of us: 'Hayek pictures human beings as following various rules and procedures which are the product of their past experience. Indeed, we have no option but to follow such rules, as they are constitutive of us – of our rationality and of the way in which we perceive the world' (Shearmur 1996a: 107).

2 Oakeshott refers to such purpose-independent rules as 'adverbial' (Oakeshott 1983: 130); for a comparison with Hayek see Barry (1994: 150).

3 In this sense, as Raphael (1998: 41) notes, the rules that constitute cricket allow us the knowledge of what is 'not cricket'.

4 Hayek and Oakeshott were aware of the similarity of their views on this issue and express agreement with each others' analysis in various places: (Hayek 1988: 37; LLL vol. 1: 125; LLL vol. 2: 112, 137; 1978: 140; Letters exchanged between Hayek and Oakeshott January 1968 to May 1968 (Friedrich A. von Hayek Papers Box no. 40, Hoover Institution Archives).

5 Perhaps the clearest development of this approach is to be found in Polanyi's work. Influenced by the gestalt psychology and study of apes by Köhler, he argues that the acquisition of habits of behaviour or skills are expressions of 'tacit' or 'personal' knowledge acquired in a particular manner; through 'subception' or 'learning without awareness' (Polanyi 1969: 143; 1946: 19). He draws a distinction between 'subsidiary' and 'focal' awareness in human perception, particularly in the 'skilful' use of 'tools', such that a pianist is focally aware of playing a particular piece of music but is only subsidiarily aware of the feel of the keys beneath his fingers (Polanyi 1958: 55–7). Both types of awareness are necessary for the successful pursuit of the activity, yet subsidiary awareness is clearly not deliberative. Polanyi notes: 'If a pianist shifts his attention from the piece he is playing to the observation of what he is doing with his fingers while playing it, he gets confused and may have to stop' (Polanyi 1958: 56). This leads him to conclude that the pursuit of focal activity is always dependent on the support of activities of which we are only subsidiarily aware. Thus I 'know' how to swim, but am unable to describe the precise muscular movement which I undertake in order to do so (Polanyi 1969: 141). Such subsidiary behaviour is, however, open to examination. Polanyi notes that: 'In

performing a skill we are therefore acting on certain premises of which we are focally ignorant, but which we know subsidiarily as part of our mastery of that skill, and which we may get to know focally by analysing the way we achieve success (or what we believe to be success) in the skill in question' (Polanyi 1958: 162). In other words we seek a functional understanding of non-deliberatively generated abilities that have emerged from an 'unconscious' process of trial and error (Polanyi 1958: 62). Polanyi notes that the tacit knowledge involved in the acquisition of skills is linked to the familiarity with the practice of the skill gained by specialists (Polanyi 1969: 188). His conception is termed 'connoisseurship' (Polanyi 1958: 54). For example, the connoisseur of wine or fine tea acquires, through experience, an 'aesthetic' recognition of classifications in his field which is not necessarily expressible or communicable. Through familiarity with his specialization the connoisseur is able to make judgements and set standards without explicitly formulating the grounds upon which they are based (Polanyi 1958: 64–5). Hayek endorses Polanyi's views on this matter (Hayek 1967: 44) and a similar argument is to be found in Hume's essay *Of the Standard of Taste*.

6 Hayek (1960: 147) rejects Mill's (1991: 77–9) argument that such social pressure is a restriction of individual liberty, for rather than acting as a barrier to self-expression, it is one of the key conditions which allow social association and which assist individual flourishing. As customs are not absolutely enforced, but rather rely on notions of propriety, they act to stabilize expectations while leaving a degree of flexibility that allows both gradual change and individual eccentricity.

7 For Popper evolution in terms of science was cultural evolution, the development of a third world of objective artefacts which could be approached in a critical manner and 'naturally selected' without the need for the death of the carrier. This aspect of cultural evolution is a Darwinian process of selection rather than a Lamarckian one of instruction by repetition (Popper 1972: 66, 144).

8 See also Oakeshott (1990: 100): 'Practices are footprints left behind by agents responding to their emergent situations.'

9 This aspect of cultural evolution is Lamarckian rather than Darwinian (Karlson 2002: 57, 71). Karlson distinguishes between evolution by natural selection, which selects by efficiency, and evolution by diffused reinforcement, which selects by imitation of the successful.

10 Galeotti sums this up well: 'Hayek's conjectural reconstruction of social spontaneity and rule formation is the following: From casual human interactions in the various spheres of social interchange, patterns emerge unintentionally. Given the human need for rules, there is a tendency to repeat those patterns as a guideline for action in future instances of similar behaviour. Then, among the number of spontaneous patterns that emerge in a given community at a given time, the most successful one has a chance to be repeated until it rules out the others' (Galeotti 1987: 171).

11 This, as with all evolutionary processes, is no guarantee that they will continue to do so in the future (Hayek 1960: 67).

12 Hayek (1984: 322; 1988: 120, 155) suggests that Smith's comments on the significance of the size of the market for the scope of the division of labour are an implicit endorsement of a population-linked theory of cultural evolution.

13 Hodgson (1991) offers the example of the Shakers whose celibacy led to the constant need to recruit new members and, in the long run, to their extinction.

14 He cites Smith approvingly on the assertion that: 'the most decisive mark of the prosperity of any country is the increase of the number of its inhabitants' (Hayek 1984: 322).

15 Similar points are raised by Kley (1994: 162), Petsoulas (2001: 63), Shearmur (1996a: 84–5), Gray (1989: 247) and Paul (1988: 259).
16 A point raised by Gray (1986: 141) and Shearmur (1996a: 86).
17 In this sense, observations of population, though not themselves a value, may carry with them certain implications. For example we may not value population growth, but be indisposed to see rapid population decline which would involve the death of large numbers.
18 For a discussion of this point see Denis (1999: 15, 32), Hodgson (1991: 79), Kley (1994: 23), Birner (1994) and Gray (1986: 52–4).
19 As Steele (1987: 181) notes group selection occurs on the level of groups within a system as well as on an individual level. Karlson (2002: 60) makes this the focus of his approach to civil society. He views change in civil society as a process of group selection: individuals join and leave groups within the broader social group and this results in individual choice affecting the order of the society as a whole.
20 Hayek also notes that it is possible to criticize a traditional practice if it is in contradiction to the other key principles upon which the order of rules rests (LLL vol. 3: 172).
21 Hayek's argument is an 'anthropology of morals' and not a 'moral philosophy' (Kukathas 1989: 202–3), the support for freedom is an empirical observation. It has been suggested that Hayek's thought runs the risk of committing the 'so-called naturalistic fallacy' which attempts to 'deduce a genuine normative state-ment from a descriptive theory' (Radnitzky 1987: 30). This, however, is not Hayek's view precisely because he provides a solely instrumental argument for the efficiency of freedom as a means to ensure wealth accumulation and the sur-vival of the population. As his critics rightly note he provides little in the way of a moral argument for the desirability of these values, but this in turn is not a critique of the details of the argument merely an assertion as to what more is required.
22 As Gissurarson (1987: 93) puts it Hayek is 'showing the empirical rather than the moral limits of benevolence'.
23 A view echoed in Popper (1966 vol. 2: 235) and Polanyi's (1951: 21) assertions that love is grounded on intimacy or familiarity, on knowledge of the other, in a sense which always renders it concrete and resistant to abstraction.
24 Brennan and Pettit (1993) refer to such approbation as an 'intangible hand' because of its fundamentally unintentional nature. We do not intend to affect a change in another's behaviour by expressing our disapproval: we merely seek to express our disapproval.

8 The evolution of law and government

1 For this reason Oakeshott notes that the office and occupant are coeval, that the opinion that a certain person should decide a case is coeval with the recognition of chiefs (Oakeshott 1990: 154).
2 Hayek, Popper and Oakeshott all accept Hume's view that government is based on opinion (Hayek 1979: 51; 1980: 60; LLL vol. 1: 55; LLL vol. 2: 13; LLL vol. 3: 33; 1978: 82, 85; 1960: 103) (Popper 1966 vol. 1: 122) (Oakeshott 1990: 156).
3 Hayek (LLL vol. 1: 73) is keen to stress that the definition of what makes a law is not that it is the will of a legislator, but rather that it rests on a notion of gradually developed general rules. The failure of legal positivism to grasp the significance of the general formal requirements of what makes a law is, for Hayek, its greatest failing (LLL vol. 3: 129).

4 And later by lawyers in the sense that a lawyer making a case appeals to pre-existing standards and seeks to shape the judgement in favour of his client (LLL vol. 1: 69).

5 Oakeshott describes the process as one of 'ruling' in the sense of the articulation of rules rather than the issuing of commands to secure a particular purpose (Oakeshott 1991: 380, 427). The articulation of law ('lex' for Oakeshott) is 'adverbial', it refers to conditions to be followed in self-chosen actions (Oakeshott 1983: 130; 1990: 263): thus civil association is a 'rule-articulated association' (Oakeshott 1990: 124).

6 Hayek compares such a notion of justice to Popper's evolutionary falsificatory model of science (LLL vol. 2: 42–3), arguing that it eliminates the illegitimate without identifying or asserting a certain end-state.

7 Thus the rules of property form part of the invisible hand argument drawing on the idea of epistemological efficiency.

8 Hence the references to Hayek as a 'system utilitarian'.

9 A point which he credits Smith with realizing (LLL vol. 2: 71).

10 We should note here that several critics of Hayek have raised the point that nothing in this analysis of general rules guarantees a respect for individual freedom (Jacobs 1999: 7). Brittan (1987: 62) highlights this point by the example of the Scottish Sabbath laws under which a general rule is restrictive of freedom of action.

11 Kley (1994: 99), however, rejects the game metaphor, arguing that, as the 'cards' are not reshuffled at the end of each round, the process lacks the common start point associated with game playing.

12 Wilhelm (1972: 170) notes that Hayek uses the conclusions of his social scientific approach to 'preclude' certain possible courses of institutional change; and Heath (1989: 109–10) is clear that the conjectural nature of the explanation leaves room for reform of the actual institutions guided by the principles of evolution and spontaneous order. As de Vlieghere puts it: 'The idea that the conjectural explanation of the origin and the success of certain institutions is useful in assessing their value in a changed society, presupposes on the one hand a functionalist confidence in spontaneous selection of the more functional institutions, and on the other hand, a possibility of rational reform based on historical research and "immanent criticism"' (De Vlieghere 1994: 294).

13 As De Crespigny (1976: 56) notes, in order truly to undermine Hayek's instrumental approach one must either disagree with the values which he believes freedom serves (rising living standards), or argue that the market does not do what he claims that it does. In this sense a cogent criticism of Hayek's work is bound to take place on his own ground.

14 Though Hayek himself does not make the distinction in quite these terms, continuing to refer to legislation in both senses but distinguishing between the nature of the law produced (LLL vol. 3: 47–8).

15 This is partly why Hayek is so distrustful of pressure groups whom he believes endanger this process by demanding minority opinions be foisted on the whole population (Hayek 1980: 116; LLL vol. 3: 11). Polanyi expresses a similar view on the role of opinion in political change: 'Public opinion is constantly making adjustments in these matters by custom and legislation' (Polanyi 1946: 56).

16 Gray (1986: 16) and Kukathas (1989: 103–4) raise the point that Hayek's immanent criticism of traditions of behaviour faces a significant problem in that he argues that traditions hold tacit knowledge of which we are unaware. The danger is that a reform will do away with some aspect of traditional behaviour which carries vital tacit knowledge. It is precisely the danger of such an ill-conceived reform that leads Hayek to advocate gradualist piecemeal reform; but

it remains nonetheless a problem from which his theory cannot escape. The flip side of this view has also been noted in that all plans for reform are inevitably grounded on established knowledge. De Vlieghere (1994: 290) notes that all constructions are based on more knowledge than the designer possesses, for example the blast furnace had to exist before the car, but the designer of a modern automobile need have no knowledge of the development of the technology necessary to create internal combustion in order to design a car. It is in this sense that Hayek believes that all progress must be based on tradition.

17 At this point we see the relevance of Hayek's pseudo-Kantian universalization test as a criterion for assessment of the rule. Universalization acts as a test of the compatibility of a proposed rule with the existing framework of rules. As a result it ensures the coherence of the framework, which in turn ensures the all-important stability of expectations. Reform of the framework has the long-term goal of improving the likelihood of satisfactory mutual adjustment.

18 Though Hayek dislikes Popper's use of the term engineering (LLL vol. 3: 193, 204 n. 50) the fundamental aspects of Popper's notion of reform appear to complement Hayek's view. For example, Popper writes of change occurring 'little by little' (Popper 1961: 75), and of 'traditions, changing and developing under the influence of critical discussion and in response to the challenge of new problems' (Popper 1989: 352).

19 Before discussing in more detail the nature of a government's role as a service provider we ought to note again that the service function is subsidiary to the lawmaking function. In the ideal constitutional model the lower house which administers service provision is subject to the generalized rules of conduct laid down by the upper house. The service functions of government must be carried out in accordance with the general rules laid down as a part of the legal framework. Which is to say that wherever possible these service functions are to be supplied generally to all citizens, and the coercive powers utilized in their implementation similarly restricted by generalized rules.

20 His specific policy statements include: a support for social insurance (Hayek 1991: 90; 1960: 286, 292), support for some form of 'workfare' scheme (Hayek 1991: 95) and an opposition to a minimum wage (Hayek 1991: 92).

9 The evolution of markets

1 We must note that in terms of economics there were significant differences between Hayek on the one hand, and Popper and Polanyi on the other, with the two latter being significantly less averse to government intervention and Keynesianism than Hayek (Magee 1973: 80–3; Prosch 1986: 186; Shearmur 1996b: 32–6). As a result our composite model of the spontaneous order approach here draws solely on the more extensive economic writings of Hayek.

2 This is consistent with Hayek's argument in the *Road to Serfdom* that links economic planning with totalitarianism. A malign unintended consequence of economic planning is the erosion of economic and political freedom.

3 See Barry (1988: 20) and Fleetwood (1996: 738–9). Caldwell (2004) suggests that Hayek turned away from technical economic theory because he saw the difficulty of applying equilibrium theory to questions of the co-ordination of knowledge.

4 Hayek deploys terms like 'ignorance', 'imperfect knowledge' and 'uncertainty' to highlight this feature of human life (Hayek 1960: 427).

5 Hayek complements this argument on the limitations of reason with the assertion that there is a species of human knowledge which is constitutionally resistant to articulation and generalization. This phenomenon, which Hayek refers to

as 'unorganized knowledge' (Hayek 1984: 214), is knowledge that does not exist in a conscious form. In a sense it may be called 'skill' rather than objective knowledge (Hayek 1980: 51).

6 Leading the spontaneous order theorists to adopt a methodological individualist approach.

7 One of the chief such questions is how humans communicate the 'value' of knowledge. Which is to say how they come to know which knowledge of circumstances is important (Hayek 1984: 257). There must exist, for the efficient exploitation of localized knowledge to come about, some incentive to acquire useful knowledge such that those in particular circumstances will exploit the relevant aspects of the circumstances in which they find themselves (Hayek 1988: 89). In brief: How do we come to know what to look for in a given situation? (Hayek 1988: 77). This becomes one of the key issues in the efficient coordination of knowledge.

8 Which Hayek, following the Scots, viewed as one of the strongest arguments in support of a system of private or 'several' property (Hayek 1988: 86–8).

9 Prices are of course based on money values and Hayek draws on Menger's conjectural history of the evolution of money from convention to explain the basis on which the price system rests (Menger 1996: 131–5; O'Driscoll 1994: 127).

10 Hayek (1991: 56) also notes the key role played in this system by the broader general rules which stabilize expectations and allow the price system to operate.

11 Hayek credits Smith with at least a partial realization of this in his focus on the negative feedback functions of prices. For Hayek Smith's invisible hand is a negative feedback knowledge communication device (LLL vol. 3: 158; 1984: 259; 1978: 63).

12 They thus parallel the state of the debate in science in Polanyi and Popper's analysis. Polanyi even refers to prices as a 'consensus' (Polanyi 1958: 208).

13 This is more accurately understood as an inter-subjectively generated objectification of value. A point raised by Hayek's pupil Shackle (1972: 220–8) and one which we have seen before in Hume's *Standard of Taste* and Popper's 'world 3'.

14 For example Kukathas (1989: 129) and Shearmur (1996a: 8).

15 There are other features that can act to pervert the information carried by prices. Among these Hayek notes: the size of the public sector (LLL vol. 1: 140); inflation (Hayek 1960: 330) and the government monopoly of the supply of money (Hayek 1978: 224; 1988: 103).

16 They are objective only in the sense that they are inter-subjectively generated and accepted.

17 Here Hayek admits a difference between his own analysis of value and that of Smith (Hayek 1984: 26). What we see here is the contribution of the Austrian school to the tradition of spontaneous order. Hayek is able to jettison the notion of an objective measure of value from his application of the spontaneous order approach and, as a consequence, provides a clearer explanation of the market price system than Smith.

Bibliography

Acton, H.B. (1972) 'Distributive Justice, the Invisible Hand and the Cunning of Reason', *Political Studies* vol. 20 no. 4, pp. 421–31.

Barry, Norman P. (1979) *Hayek's Social and Economic Philosophy*, London: Macmillan.

—— (1982) 'The Tradition of Spontaneous Order', *Literature of Liberty* vol. 5, pp. 7–58.

—— (1986) *On Classical Liberalism and Libertarianism*, London: Macmillan.

—— (1988) *The Invisible Hand in Economics and Politics*, London: Institute of Economic Affairs Hobart Paper 111.

—— (1994) 'The Road to Freedom: Hayek's Social and Economic Philosophy', in J. Birner and R. van Zijp (eds) *Hayek, Co-ordination and Evolution; His Legacy in Philosophy, Politics, Economics and the History of Ideas*, London: Routledge, pp. 141–63.

Baumgarth, William P. (1978) 'Hayek and Political Order: the Rule of Law', *Journal of Libertarian Studies* vol. 2 no. 1, pp. 11–28.

Becker, James F. (1961) 'Adam Smith's Theory of Social Science', *Southern Economic Journal* vol. 28 no. 1, pp. 13–21.

Berry, Christopher J. (1982) *Hume, Hegel and Human Nature*, The Hague: Martinus Nijhoff.

—— (1989) 'Adam Smith: Commerce, Liberty and Modernity', in P. Gilmour (ed.) *Philosophers of the Enlightenment*, Edinburgh: Edinburgh University Press, pp. 113–32.

—— (1994) *The Idea of Luxury: A Conceptual and Historical Investigation*, Cambridge: Cambridge University Press.

—— (1997) *Social Theory of the Scottish Enlightenment*, Edinburgh: Edinburgh University Press.

Birner, Jack (1994) 'Hayek's Grand Research Programme', in J. Birner and R. van Zijp (eds) *Hayek, Co-ordination and Evolution: His Legacy in Philosophy, Politics, Economics and the History of Ideas*, London: Routledge, pp. 1–21.

Bishop, John D. (1995) 'Adam Smith's Invisible Hand Argument', *Journal of Business Ethics* vol. 14, pp. 165–80.

Boudon, Raymond (1982) *The Unintended Consequences of Social Action*, London: Macmillan.

Bowles, Paul (1985) 'The Origin of Property and the Development of Scottish Historical Science', *Journal of the History of Ideas* vol. 46, pp. 197–209.

Brennan, G. and Pettit, P. (1993) 'Hands Invisible and Intangible', *Synthese* vol. 94, pp. 191–225.

Brewer, John D. (1986) 'Adam Ferguson and the Theme of Exploitation', *The British Journal of Sociology* vol. 37, pp. 461–78.

Brittan, Samuel (1987) 'Hayek, Freedom, and Interest Groups', in E. Butler and M. Pirie (eds) *Hayek on the Fabric of Human Society*, London: Adam Smith Institute, pp. 47–75.

Broadie, Alexander (2001) *The Scottish Enlightenment*, Edinburgh: Birlinn.

Bronk, Richard (1998) *Progress and the Invisible Hand: The Philosophy and Economics of Human Advance*, London: Warner Books.

Brown, Maurice (1988) *Adam Smith's Economics; Its Place in the Development of Economic Thought*, London: Croom Helm.

Bryson, Gladys (1968) *Man and Society: The Scottish Inquiry of the Eighteenth Century*, New York: Augustus M. Kelley.

Buchanan, James M. (1977) *Freedom in Constitutional Contract: Perspectives of a Political Economist*, College Station, TX: Texas A & M University Press.

Burke, Peter (1985) *Vico*, Oxford: Oxford University Press.

Butler, Eamonn (1983) *Hayek: his Contribution to the Political and Economic Thought of our Time*, London: Temple Smith.

—— (1987) 'Hayek on the Evolution of Morality', in E. Butler and M. Pirie (eds) *Hayek on the Fabric of Human Society*, London: Adam Smith Institute, pp. 113–26.

Caldwell, Bruce J. (1994) 'Hayek's Scientific Subjectivism', *Economics and Philosophy* vol. 10, pp. 305–13.

—— (2004) *Hayek's Challenge: An Intellectual Biography of F.A. Hayek*, Chicago: The University of Chicago Press.

Camic, Charles (1983) *Experience and Enlightenment: Socialization for Cultural Change in Eighteenth-Century Scotland*, Edinburgh: Edinburgh University Press.

Campbell, T.D. (1975) 'Scientific Explanation and Ethical Judgement in the Moral Sentiments', in Andrew S. Skinner and Thomas Wilson (eds) *Essays on Adam Smith*, Oxford: Clarendon, pp. 68–82.

—— (1977) 'Adam Smith and Natural Liberty', *Political Studies* vol. 25 no. 4, pp. 523–34.

Campbell, T.D and Ross, I.S. (1981) 'The Utilitarianism of Adam Smith's Policy Advice', *Journal of the History of Ideas* vol. 42, pp. 73–92.

Carabelli, Anna and De Vecchi, Nicolo (2001) 'Individuals, Public Institutions and Knowledge', in Pier Luigi Porta, Roberto Scazzieri and Andrew Skinner (eds) *Knowledge, Social Institutions and the Division of Labour*, Cheltenham: Edward Elgar, pp. 229–48.

Chitnis, A. (1976) *The Scottish Enlightenment: A Social History*, London: Croom Helm.

Clark, Henry C. (1992) 'Conversation and Moderate Virtue in Adam Smith's Theory of Moral Sentiments', *Review of Politics* vol. 54, pp. 185–210.

—— (1993) 'Women and Humanity in Scottish Enlightenment Social Thought: The Case of Adam Smith', *Historical Reflections* vol. 19, no. 3, pp. 335–61.

Coats, A.W. (1975) 'Adam Smith and the Mercantile System', in Andrew S. Skinner and Thomas Wilson (eds) *Essays on Adam Smith*, Oxford: Clarendon, pp. 219–36.

Coleman, Samuel (1968) 'Is There Reason in Tradition?', in P. King and B.C. Parekh (eds) *Politics and Experience: Essays Presented to Professor Michael Oakeshott on the Occasion of his Retirement*, Cambridge: Cambridge University Press, pp. 239–82.

Condren, Conal (1985) *The Status and Appraisal of Classic Texts: An Essay on Political Theory, its Inheritance, and the History of Ideas*, Princeton: Princeton University Press.

Connin, Lawrence J. (1990) 'Hayek, Liberalism and Social Knowledge', *Canadian Journal of Political Science* vol. 23 no. 2, pp. 297–315.

Cosmides, Leda and Tooby, John (1994) 'Better than Rational: Evolutionary Psychology and the Invisible Hand', *American Economic Association Papers and Proceedings* (May), pp. 327–32.

Cropsey, Joseph (1957) *Polity and Economy: An Interpretation of the Principles of Adam Smith*, The Hague: Martinus Nijhoff.

Danford, John W. (1990) 'Hume's History and the Parameters of Economic Development', in Nicholas Capaldi and Donald W. Livingston (eds) *Liberty in Hume's History of England*, Dordrecht: Kluwer Academic, pp. 155–94.

Darwin, Charles (1998) [1859] *The Origin of Species*, Ware, Hertfordshire: Wordsworth Editions.

Davis, J.B. (1983) 'Smith's Invisible Hand and Hegel's Cunning of Reason', in John Cunningham Wood (ed.) *Adam Smith: Critical Assessments*, vol. 6, London: Croom Helm, pp. 300–20.

Dawkins, Richard (1986) *The Blind Watchmaker*, Harlow: Longman Scientific and Technical.

De Crespigny, Anthony (1976) 'F.A. Hayek: Freedom for Progress', in Anthony de Crespigny and Kenneth Minogue (eds) *Contemporary Political Philosophers*, London: Methuen, pp. 49–66.

Denis, Andy (1999) 'Friedrich Hayek: A Panglossian Evolutionary Theorist', http://www.city.ac.uk/~rc369/research/Hayek.html (accessed 7/2/00 1420).

De Vlieghere, Martin (1994) 'A Reappraisal of Friedrich A. Hayek's Cultural Evolutionism', *Economics and Philosophy* vol. 10, pp. 285–304.

Dickey, Laurence (1986) 'Historicizing the "Adam Smith Problem": Conceptual, Historiographical, and Textual Issues', *Journal of Modern History* vol. 58, pp. 579–609.

Dickinson, H.T. (1975) 'The Politics of Bernard Mandeville', in I. Primer (ed.) *Mandeville Studies: New Explorations in the Art and Thought of Dr. Bernard Mandeville (1630–1733)*, The Hague: Martinus Nijhof, pp. 80–97.

Dunbar, J. (1995) [1781] *Essays on the History of Mankind in Rude and Cultivated Ages*, 2nd edn, Bristol: Thoemmes Press reprint.

Dwyer, John (1998) *The Age of the Passions: An Interpretation of Adam Smith and Scottish Enlightenment Culture*, East Linton: Tuckwell Press.

Ebenstein, Alan (2003) *The Mind of Friedrich Hayek*, New York: Palgrave Macmillan.

Elster, Jon (1989a) *The Cement of Society: A Study of Social Order*, Cambridge: Cambridge University Press.

—— (1989b) *Nuts and Bolts for the Social Sciences*, Cambridge: Cambridge University Press.

Evensky, Jerry (1993) 'Ethics and the Invisible Hand', *Journal of Economic Perspectives* vol. 7 no. 2, pp. 197–205.

Ferguson, Adam (1973) [1792] *Principles of Moral and Political Science*, 2 vols. New York: AMS Press.

—— (1994) [1769] *Institutes of Moral Philosophy*, London: Routledge/Thoemmes Press.

—— (1995) [1767] *An Essay on the History of Civil Society*, ed. Fania Oz-Salzberger, Cambridge: Cambridge University Press.

Fleetwood, Steve (1996) 'Order Without Equilibrium: a Critical Realist

Interpretation of Hayek's Notion of Spontaneous Order', *Cambridge Journal of Economics* vol. 20, pp. 729–47.

Flew, Antony (1985) *Thinking About Social Thinking: The Philosophy of the Social Sciences*, Oxford: Basil Blackwell.

—— (1986) *David Hume: Philosopher of Moral Science*, Oxford: Basil Blackwell.

—— (1987) 'Social Science: Making Visible the Invisible Hand', *Journal of Libertarian Studies* vol. 8 no. 2, pp. 197–211.

Forbes, Duncan (1954) 'Scientific Whiggism: Adam Smith and John Millar', *The Cambridge Journal* vol. 7 (August), pp. 643–70.

—— (1975) 'Sceptical Whiggism, Commerce and Liberty', in A. Skinner and T. Wilson (eds) *Essays on Adam Smith*, Oxford: Clarendon, pp. 179–201.

Galeotti, Anna E. (1987) 'Individualism, Social Rules, Tradition: The Case of Friedrich A. Hayek', *Political Theory* vol. 15 no. 2, pp. 163–81.

Gay, Peter (1967) *The Enlightenment: An Interpretation, Volume I The Rise of Modern Paganism*, London: Weidenfeld & Nicolson.

—— (1969) *The Enlightenment: An Interpretation, Volume II The Science of Freedom*, London: Weidenfeld & Nicolson.

Gee, J.M.A. (1968) 'Adam Smith's Social Welfare Function', *Scottish Journal of Political Economy* vol. 15, pp. 283–99.

Gellner, Ernest (1994) *Conditions of Liberty: Civil Society and its Rivals*, Harmondsworth: Penguin.

Gissurarson, Hannes H. (1987) *Hayek's Conservative Liberalism*, London: Garland Press.

Goldsmith, M.M. (1985) *Private Vices, Public Benefits: Bernard Mandeville's Social and Political Thought*, Cambridge: Cambridge University Press.

—— (1988) 'Regulating Anew the Moral and Political Sentiment of Mankind: Bernard Mandeville and the Scottish Enlightenment', *Journal of the History of Ideas* vol. 46, pp. 197–209.

—— (1994) 'Liberty, Virtue, and the Rule of Law, 1689–1770', in D. Wootton (ed.) *Republicanism, Liberty and Civil Society 1649–1776*, Stanford: Stanford University Press, pp. 197–232.

Graham, David and Clarke, Peter (1986) *The New Enlightenment: The Rebirth of Liberalism*, London: Macmillan.

Grant, Robert (1990) *Oakeshott*, London: The Claridge Press.

Gray, Sir Alexander (1931) *The Development of Economic Doctrine: An Introductory Survey*, London: Longmans, Green and Co.

Gray, John (1986) *Hayek on Liberty*, 2nd edn, Oxford: Basil Blackwell.

—— (1989) *Liberalisms: Essays in Political Philosophy*, London: Routledge.

—— (1990) 'Hayek, the Scottish School, and Contemporary Economics', in D. Mair (ed.) *The Scottish Contribution to Modern Economic Thought*, Aberdeen: Aberdeen University Press, pp. 249–62.

Green, David G. (1987) *The New Right: The Counter-Revolution in Political, Economic and Social Thought*, Brighton: Wheatsheaf Books.

Griswold, Charles L. Jr. (1996) 'Nature and Philosophy: Adam Smith on Stoicism, Aesthetic Reconciliation, and Imagination', *Man and World* vol. 29, pp. 187–213.

—— (1999) *Adam Smith and the Virtues of Enlightenment*, Cambridge: Cambridge University Press.

Haakonssen, Knud (1981) *The Science of a Legislator*, Cambridge: Cambridge University Press.

—— (1982) 'What Might Properly be Called Natural Jurisprudence', in R.H. Campbell and A.S. Skinner (eds) *The Origins and Nature of the Scottish Enlightenment*, Edinburgh: John Donald, pp. 205–25.

—— (1990) 'Natural Law and Moral Realism: The Scottish Synthesis', in M.A. Stewart (ed.) *Studies in the Philosophy of the Scottish Enlightenment*, Oxford: Clarendon, pp. 61–85.

Hamowy, Ronald (1987) *The Scottish Enlightenment and the Theory of Spontaneous Order*, Carbondale: Southern Illinois University Press.

Harpham, Edward J. (1983) 'Liberalism, Civic Humanism, and the Case of Adam Smith', *The American Political Science Review* vol. 78 no. 3, pp. 764–74.

Hayek, F.A. (1960) *The Constitution of Liberty*, London: Routledge.

—— (1967) *Studies in Philosophy, Politics and Economics*, London: Routledge and Kegan Paul.

—— (1976) [1952] *The Sensory Order*, London: Routledge and Kegan Paul.

—— (1979) *The Counter-Revolution of Science: Studies on the Abuse of Reason*, Indianapolis: Liberty Fund.

—— (1978) *New Studies in Philosophy, Politics, Economics and the History of Ideas*, London: Routledge and Kegan Paul.

—— (1980) [1948] *Individualism and Economic Order*, Chicago: University of Chicago Press.

—— (1984) *The Essence of Hayek*, ed. Chiaki Nishiyama and Kurt R. Leube, Stanford: Hoover Institution Press.

—— (1988) *The Fatal Conceit: The Errors of Socialism*, ed. W.W. Bartley III, London: Routledge.

—— (1991) [1944] *The Road to Serfdom*, London: Routledge.

—— (1993) [1973–82] *Law, Legislation and Liberty*, 3 vols. London: Routledge.

—— (1994) *Hayek on Hayek: An Autobiographical Dialogue*, Stephen Kresge and Leif Wenar (eds), London: Routledge.

—— Friedrich A. von Hayek Papers, Boxes 40–3, Hoover Institution Archives.

Heath, Eugene (1989) 'How to Understand Liberalism as Gardening: Galeotti on Hayek', *Political Theory* vol. 17 no. 1, pp. 107–13.

—— (1995) 'The Commerce of Sympathy: Adam Smith on the Emergence of Morals', *Journal of the History of Philosophy* 33, pp. 447–66.

Heilbroner, R.L. (1975) 'The Paradox of Progress: Decline and Decay in the Wealth of Nations', in Andrew S. Skinner and Thomas Wilson (eds) *Essays on Adam Smith*, Oxford: Clarendon, pp. 524–39.

Hetherington, Norriss S. (1983) 'Isaac Newton's Influence on Adam Smith's Natural Laws in Economics', *Journal of the History of Ideas* vol. 44, pp. 497–505.

Hill, Lisa (1997) 'Adam Ferguson and the Paradox of Progress and Decline', *History of Political Thought* vol. 18 no. 4, pp. 677–706.

Hirschman, Albert O. (1977) *The Passions and the Interests: Political Arguments for Capitalism Before its Triumph*, Princeton, NJ: Princeton University Press.

Hodgson, Geoffrey M. (1991) 'Hayek's Theory of Cultural Evolution: An Evaluation in the Light of Vanberg's Critique', *Economics and Philosophy* vol. 7, pp. 67–82.

Hollander, Samuel (1973) *The Economics of Adam Smith*, London: Heinemann Educational.

—— (1975) 'On the Role of Utility and Demand in the Wealth of Nations', in Andrew S. Skinner and Thomas Wilson (eds) *Essays on Adam Smith*, Oxford: Clarendon, pp. 313–23.

Hont, Istvan (1987) 'The Language of Sociability and Commerce: Samuel Pufendorf and the Theoretical Foundations of the "Four Stages Theory"', in Anthony Pagden (ed.) *The Language of Political Theory in Early-Modern Europe*, Cambridge: Cambridge University Press, pp. 253–76.

Hope, V.M. (1989) *Virtue by Consensus: The Moral Philosophy of Hutcheson, Hume and Adam Smith*, Oxford: Clarendon.

Höpfl, H.M. (1978) 'From Savage to Scotsman: Conjectural History in the Scottish Enlightenment, *Journal of British Studies* vol. 17 no. 2, pp. 19–40.

Horne, Thomas A. (1978) *The Social Thought of Bernard Mandeville: Virtue and Commerce in Early Eighteenth-Century England*, London: Macmillan.

Hume, David (1975) [1777] *Enquiries Concerning Human Understanding and Concerning the Principles of Morals*, 3rd edition, ed. L.A. Selby-Bigge, rev. P.H. Nidditch, Oxford: Clarendon Press.

—— (1978) [1739] *A Treatise of Human Nature*, 2nd edition, ed. L.A. Selby-Bigge, rev. P.H. Nidditch, Oxford: Clarendon.

—— (1983) [1778] *The History of England*, 6 volumes, Indianapolis: Liberty Fund.

—— (1985) [1777] *Essays Moral, Political, and Literary*, ed. Eugene F. Miller, Indianapolis: Liberty Fund.

Hundert, E.G. (1994) *The Enlightenment's Fable: Bernard Mandeville and the Discovery of Society*, Cambridge: Cambridge University Press.

Hutchison, Terence (1990a) 'History and Political Economy in Scotland: Alternative Inquiries and Scottish Ascendency', in D. Mair (ed.) *The Scottish Contribution to Modern Economic Thought*, Aberdeen: Aberdeen University Press, pp. 61–80.

—— (1990b) 'Adam Smith and the Wealth of Nations', in D. Mair (ed.) *The Scottish Contribution to Modern Economic Thought*, Aberdeen: Aberdeen University Press, pp. 81–102.

Jacobs, Struan (1998) 'Michael Polanyi and Spontaneous Order 1941–1951', *Tradition and Discovery: The Polanyi Society Periodical* vol. 24 no. 2, pp. 14–28.

—— (1999) 'Classical and Conservative Liberalism: Burke, Hayek, Polanyi and Others', *Tradition and Discovery: The Polanyi Society Periodical* vol. 26 no. 1, pp. 5–15.

—— (2000) 'Spontaneous Order: Michael Polanyi and Friedrich Hayek', *Critical Review of International Social and Political Philosophy* vol. 3 no. 4, pp. 49–67.

Jones, Peter (1990) 'On Reading Hume's History of Liberty', in Nicholas Capaldi and Donald W. Livingston (eds) *Liberty in Hume's History of England*, Dordrecht: Kluwer Academic, pp. 1–23.

Kames, Lord (1751) *Essays on the Principles of Morality and Natural Religion*, 3rd edn. (corrected and improved 1779).

—— (1774) *Sketches of the History of Man*, 3rd edn, Dublin.

—— (1776) [1758] *Historical Law Tracts*, 3rd edn, Edinburgh.

Karlson, Nils (2002) *The State of State: Invisible Hands in Politics and Civil Society*, New Brunswick and London: Transaction Publishers.

Kerkhof, Bert (1995) 'A Fatal Attraction? Smith's "Theory of Moral Sentiments" and Mandeville's "Fable"', *History of Political Thought* vol. 16 no. 2, pp. 219–33.

Kettler, David (1965) *The Social and Political Thought of Adam Ferguson*, Columbus: Ohio State University Press.

—— (1977) 'History and Theory in Ferguson's Essay on the History of Civil Society', *Political Theory* vol. 5 no. 4, pp. 437–60.

Kleer, Richard A. (1995) 'Final Causes in Adam Smith's Theory of Moral Senti-ments', *Journal of the History of Philosophy* vol. 33, pp. 275–300.

Klein, Daniel B. (1997) 'Convention, Social Order, and the Two Co-ordinations, *Constitutional Political Economy* vol. 8 no. 4, pp. 319–35.

Kley, Roland (1994) *Hayek's Social and Political Thought*, Oxford: Clarendon.

Kukathas, Chandran (1989) *Hayek and Modern Liberalism*, Oxford: Clarendon Press.

Legutko, Ryszand (1997) 'Was Hayek an Instrumentalist?', *Critical Review* vol. 11 no. 1, pp. 145–64.

Lehmann, W.C. (1930) *Adam Ferguson and the Beginnings of Modern Sociology*, New York: Columbia University Press.

Lessnoff, Michael H. (1974) *The Structure of Social Science: A Philosophical Introduction*, London: George Allen & Unwin.

—— (1999) *Political Philosophers of the Twentieth Century*, Oxford: Blackwell.

Livingston, Donald W. (1990) 'Hume's Historical Conception of Liberty', in Nicholas Capaldi and Donald W. Livingston (eds) *Liberty in Hume's History of England*, pp. 105–53.

Lopreato, Joseph (1984) *Human Nature and Biocultural Evolution*, Boston: Allen and Unwin.

Macfie, Alec (1967) *The Individual in Society: Papers on Adam Smith*, London: George Allen and Unwin.

—— (1971) 'The Invisible Hand of Jupiter', *Journal of the History of Ideas* vol. 32, pp. 595–9.

—— (1990) 'The Scottish Tradition in Economic Thought', in D. Mair (ed.) *The Scottish Contribution to Modern Economic Thought*, Aberdeen: Aberdeen University Press, pp. 1–18.

MacRae, Donald G. (1969) 'Adam Ferguson', in Timothy Raison (ed.) *The Founding Fathers of Social Science*, Harmondsworth: Penguin, pp. 17–26.

Magee, Bryan (1973) *Popper*, London: Fortuna/Collins.

Mandeville, Bernard (1988) *The Fable of the Bees or Private Vices, Publick Benefits*, 2 vols. ed. F.B. Kaye, Indianapolis: Liberty Fund.

Martin, D.A. (1983) 'Economics as Ideology: On Making the "Invisible Hand" Invisible', in John Cunningham Wood (ed.) *Adam Smith: Critical Assessments* vol. 7, London: Croom Helm, pp. 123–37.

McDowell, Gary L. (1983) 'Commerce, Virtue, and Politics: Adam Ferguson's Con-stitutionalism', *Review of Politics* vol. 45, pp. 536–52.

Meek, Ronald L. (1971) 'Smith, Turgot, and the "Four Stages" Theory', *History of Political Economy* vol. 3, pp. 9–27.

—— (1976) *Social Science and the Ignoble Savage*, Cambridge: Cambridge University Press.

Megill, A.D. (1975) 'Theory and Experience in Adam Smith', *Journal of the History of Ideas* vol. 36, pp. 79–94.

Menger, Karl (1996) *Investigations into the Method of the Social Sciences*, trans. Francis J. Nock, Grove City, PA: Libertarian Press.

Merton, Robert K. (1957) *Social Theory and Social Structure*, New York: The Free Press.

—— (1976) *Sociological Ambivalence and Other Essays*, New York: The Free Press.

Mill, John Stuart (1991) *On Liberty and Other Essays*, ed. John Gray, Oxford: Oxford University Press.

Millar, John (1812) [1803] *An Historical View of the English Government*, 4 vols, London.

—— (1990) [1806] *The Origin of the Distinction of Ranks*, Bristol: Thoemmes Press.

 Minogue, Kenneth (1985) *Alien Powers: The Pure Theory of Ideology*, London: Weidenfeld & Nicolson.

—— (1987) 'Hayek and Conservatism: Beatrice and Benedick?' in E. Butler and M. Pirie (eds) *Hayek on the Fabric of Human Society*, London: Adam Smith Institute.

Mises, Ludwig von (1951) *Socialism: An Economic and Sociological Analysis*, trans. J. Kahane, London: Jonathan Cape.

—— (1978) *Liberalism: A Socio-Economic Exposition*, 2nd ed., trans. Ralph Raico, ed. Arthur Goddard, Kansas City: Sheed Andrews and McMeel.

Mizuta, Hiroshi (1975) 'Moral Philosophy and Civil Society', in Andrew S. Skinner and Thomas Wilson (eds) *Essays on Adam Smith*, Oxford: Clarendon, pp. 114–31.

Nozick, Robert (1974) *Anarchy, State, and Utopia*, Oxford: Blackwell.

—— (1981) *Philosophical Explanations*, Oxford: Clarendon.

—— (1994) 'Invisible-Hand Explanations', *American Economic Association Papers and Proceedings* (May), pp. 314–18.

Oakeshott, Michael (1983) *On History and Other Essays*, Oxford: Basil Blackwell.

—— (1990) *On Human Conduct*, Oxford: Clarendon.

—— (1991) *Rationalism in Politics and Other Essays*, Indianapolis: Liberty Fund.

O'Driscoll, Gerald P. (1977) *Economics as a Coordination Problem: The Contribution of Friedrich A. Hayek*, Kansas City: Sheed Andrews and McMeel.

—— (1994) 'An Evolutionary Approach to Banking and Money', in J. Birner and R. van Zijp (eds) *Hayek, Co-ordination and Evolution: His Legacy in Philosophy, Politics, Economics and the History of Ideas*, London: Routledge, pp. 126–37.

O'Hear, Anthony (1992) 'Criticism and Tradition in Popper, Oakeshott and Hayek', *Journal of Applied Philosophy* vol. 9 no. 1, pp. 65–75.

Otteson, James R. (2002) *Adam Smith's Marketplace of Life*, Cambridge: Cambridge University Press.

Paul, Ellen Frankel (1977) 'Adam Smith: A Reappraisal', *Journal of Libertarian Studies* vol. 1 no. 4, pp. 289–306.

—— (1988) 'Liberalism, Unintended Orders and Evolutionism, *Political Studies* vol. 36, pp. 251–72.

Petsoulas, Christina (2001) *Hayek's Liberalism and its Origins: His Idea of Spontaneous Order and the Scottish Enlightenment*, London: Routledge.

Phillipson, N.T. and Mitchison, R. (1996) 'Introduction' to *Scotland in the Age of Improvement*, Edinburgh: Edinburgh University Press, pp. 1–4.

Pocock, J.G.A. (1975) *The Machiavellian Moment: Florentine Political Thought and the Atlantic Republican Tradition*, Princeton: Princeton University Press.

Polanyi, Michael (1946) *Science, Faith and Society*, London: Geoffrey Cumberlege/ Oxford University Press.

—— (1951) *The Logic of Liberty*, London: Routledge.

—— (1958) *Personal Knowledge: Towards a Post-Critical Philosophy*, London: Routledge and Kegan Paul.

—— (1969) *Knowing and Being*, ed. Marjorie Grene, London: Routledge.

Popper, Karl R. (1959) *The Logic of Scientific Discovery*, London: Hutchinson & Co.

—— (1961) *The Poverty of Historicism*, 2nd edn., London: Routledge and Kegan Paul.

—— (1966) *The Open Society and its Enemies*, 2 vols, London: Routledge.

—— (1972) *Objective Knowledge: An Evolutionary Approach*, London: Routledge.

—— (1989) *Conjectures and Refutations*, London: Routledge.

—— (1994) *The Myth of the Framework: In Defence of Science and Rationality*, ed. M.A. Notturno, London: Routledge.

Porta, Pier Luigi and Scazzieri, Roberto (2001) 'Coordination, Connecting Principles and Social Knowledge: An Introductory Essay', in Pier Luigi Porta, Roberto Scazzieri and Andrew Skinner (eds) *Knowledge, Social Institutions and the Division of Labour*, Cheltenham: Edward Elgar, pp. 1–32.

Postrel, Virginia (1998) *The Future and its Enemies: The Growing Conflict Over Creativity, Enterprise, and Progress*, New York: The Free Press.

Prosch, Harry (1986) *Michael Polanyi: A Critical Exposition*, Albany, NY: State University of New York Press.

Radnitzky, Gerald (1987) 'The Constitutional Protection of Liberty', in E. Butler and M. Pirie (eds) *Hayek on the Fabric of Human Society*, London: Adam Smith Institute.

Raphael, D.D. (1975) 'The Impartial Spectator', in Andrew S. Skinner and Thomas Wilson (eds) *Essays on Adam Smith*, Oxford: Clarendon Press, pp. 83–99.

—— (1979) 'Adam Smith: Philosophy, Science and Social Science', in S.C. Brown (ed.) *Philosophers of the Enlightenment*, Brighton: Harvester Press, pp. 77–93.

Raphael, Frederic (1998) *Karl Popper, Historicism and its Poverty*, London: Phoenix Paperbacks.

Rashid, Salim (1998) *The Myth of Adam Smith*, Cheltenham: Edward Elgar.

Recktenwald, Horst Claus (1990) 'An Adam Smith Rennaisance anno 1976? The Bicentary Output – A Reappraisal of his Scholarship', in D. Mair (ed.) *The Scottish Contribution to Modern Economic Thought*, Aberdeen: Aberdeen University Press, pp. 103–34.

Redman, D.A. (1993) 'Adam Smith and Isaac Newton', *Scottish Journal of Political Economy* vol. 40, pp. 210–30.

Reisman, D.A. (1976) *Adam Smith's Sociological Economics*, London: Croom Helm.

Rendall, Jane (1978) *The Origins of the Scottish Enlightenment*, London: Macmillan.

Robertson, John (1985) *The Scottish Enlightenment and the Militia Issue*, Edinburgh: John Donald.

Robertson, William (1769) 'A View of the Progress of Society in Europe', in *Works*, ed. D. Stewart, in one volume, Edinburgh 1840.

—— (1777) 'The History of America', in *Works*, ed. D. Stewart, in one volume, Edinburgh 1840.

Rosenberg, Nathan (1963) 'Mandeville and Laissez-Faire', *Journal of the History of Ideas* vol. 24, pp. 183–96.

—— (1965) 'Adam Smith on the Division of Labour: Two Views or One?', *Economica* vol. 32 nos 125–8, pp. 127–39.

—— (1976) 'Another Advantage of the Division of Labour', *Journal of Political Economy* vol. 84 no. 4.1, pp. 861–68.

—— (1990a) 'Adam Smith as a Social Critic', *Royal Bank of Scotland Review*, no. 166 June, pp. 17–33.

—— (1990b) 'Adam Smith and the Stock of Moral Capital', *History of Political Economy* vol 22 no. 1, pp. 1–17.

Ross, Ian Simpson (1995) *The Life of Adam Smith*, Oxford: Clarendon.

Rothschild, Emma (1994) 'Adam Smith and the Invisible Hand', *American Economic Association Papers and Proceedings* (May), pp. 319–22.

—— (2001) *Economic Sentiments: Adam Smith, Condorcet, and the Enlightenment*, Cambridge, MA: Harvard University Press.

Schneider, Louis (1967) *The Scottish Moralists on Human Nature and Society*, Chicago: University of Chicago Press.

—— (1980) 'Introduction' to *An Essay on the History of Civil Society*, New Brunswick: Transaction Publishers, pp. v–xxviii.

Shackle, G.L.S. (1972) *Epistemics and Economics: A Critique of Economic Doctrines*, Cambridge: Cambridge University Press.

Shearmur, Jeremy (1996a) *Hayek and After: Hayekian Liberalism as a Research Programme*, London: Routledge.

—— (1996b) *The Political Thought of Karl Popper*, London: Routledge.

Sher, R. (1989) 'Adam Ferguson, Adam Smith and the Problem of National Defence', *Journal of Modern History* vol. 61 (June), pp. 240–68.

—— (1994) 'From Troglodytes to Americans: Montesquieu and the Scottish Enlightenment on Liberty, Virtue, and Commerce', in D.Wootton (ed.) *Republicanism, Liberty, and Commercial Society 1649–1776*, Stanford: Stanford University Press, pp. 368–402.

Shils, Edward (1997) *The Virtue of Civility: Selected Essays on Liberalism, Tradition and Civil Society*, Indianapolis: Liberty Fund.

Skinner, Andrew S. (1965) 'Economic and History – The Scottish Enlightenment', *Scottish Journal of Political Economy* vol. 12, pp. 1–22.

—— (1967) 'Natural History in the Age of Adam Smith', *Political Studies* vol. 15 no. 1, pp. 32–48.

—— (1974) 'Adam Smith, Science and the Role of the Imagination', in W.B. Todd (ed.) *Hume and the Enlightenment: Essays Presented to Ernest Campbell Mossner*, Edinburgh: Edinburgh University Press, pp. 164–88.

—— (1990) 'Adam Smith and Economic Liberalism', in D. Mair (ed.) *The Scottish Contribution to Modern Economic Thought*, Aberdeen: Aberdeen University Press, pp. 135–54.

—— (1996) *A System of Social Science: Papers Relating to Adam Smith*, 2nd edn, Oxford: Clarendon Press.

Skinner, Quentin (1969) 'Meaning and Understanding in the History of Ideas', *History and Theory* vol. 8, pp. 3–53.

Smith, Adam (1976) [1776] *An Inquiry into the Nature and Causes of the Wealth of Nations*, ed. R.H. Campbell, A.S. Skinner and W.B. Todd, Oxford: Oxford University Press.

—— (1976b) [1759] *The Theory of Moral Sentiments*, ed. D.D. Raphael and A.L. Macfie, Oxford: Oxford University Press.

—— (1978) *Lectures on Jurisprudence*, ed. R.L. Meek, D.D. Raphael and P.G. Stein, Oxford: Oxford University Press.

—— (1980) [1795] *Essays on Philosophical Subjects*, ed. W.P.D. Wightman, Oxford: Oxford University Press.

—— (1983) *Lectures on Rhetoric and Belles Lettres*, ed. J.C. Bryce, Oxford: Oxford University Press.

—— (1987) *The Correspondence of Adam Smith*, ed. Ernest Campbell Mossner and Ian Simpson Ross, Indianapolis: Liberty Fund.

Smith, Craig (Forthcoming 2005/6) 'Adam Smith on Progress and Knowledge', in E. Schliesser and L. Montes (eds) *New Voices on Adam Smith*, London: Routledge.

—— (Forthcoming 2006) 'Hayek and Spontaneous Order', in N. Barry (ed.) *The Elgar Companion to Hayekian Economics*, Cheltenham: Edward Elgar.

Smith, Vernon (1997) *The Two Faces of Adam Smith*, Southern Economic Association Distinguished Guest Lecture, Atlanta November 21 1997, Tuscon: University of Arizona Economic Science Laboratory.

Spengler, J.J. (1977) 'Adam Smith on Human Capital', *American Economic Review* vol. 67 (Feb.), pp. 32–6.

—— (1983) 'Adam Smith on Population Growth and Economic Development', in John Cunningham Wood (ed.) *Adam Smith: Critical Assessments* vol. 3, London: Croom Helm, pp. 395–406.

Steele, David Ramsay (1987) 'Hayek's Theory of Cultural Group Selection', *Journal of Libertarian Studies* vol. 8 no. 2, pp. 171–95.

Stein, Peter (1996) 'Law and Society in Scottish Thought', in N.T. Phillipson and R. Mitchison (eds) *Scotland in the Age of Improvement*, Edinburgh: Edinburgh University Press, pp. 148–68.

Stewart, Dugald (1793) 'Account of the Life and Writings of Adam Smith LL.D', in W.P.D. Wightman (ed.) Adam Smith's *Essays on Philosophical Subjects*, Oxford: Oxford University Press (1980), pp. 269–351.

Stewart, M.A. (1994) 'Hume's Historical View of Miracles', in M.A. Stewart and John P. Wright (eds) *Hume and Hume's Connexions*, Edinburgh: Edinburgh University Press, pp. 171–200.

Stuart, G. (1768) *Historical Dissertation concerning the Antiquity of the English Constitution*, Edinburgh.

—— (1995) [1792] *A View of Society in Europe in its Progress from Rudeness to Refinement*, 2nd edn, Bristol: Thoemmes Press.

Sufrin, Sidney C. (1961) 'Some Reflections on Hayek's Constitution of Liberty', *Ethics* vol. 71, pp. 201–4.

Sugden, Robert (1989) 'Spontaneous Order', *Journal of Economic Perspectives* vol. 3 no. 4, Fall, pp. 85–97.

Swingewood, Alan (1970) 'Origins of Sociology: the Case of the Scottish Enlightenment', *British Journal of Sociology* vol. 21, pp. 164–80.

Teichgraeber III, Richard F. (1981) 'Rethinking Das Adam Smith Problem', *Journal of British Studies* vol. 20, pp. 106–23.

—— (1986) *Free Trade and Moral Philosophy: Rethinking the Sources of Adam Smith's Wealth of Nations*, Durham, NC: Duke University Press.

Ullman-Margalit, Edna (1977) *The Emergence of Norms*, Oxford: Clarendon.

—— (1978) 'Invisible-Hand Explanations', *Synthese* vol. 39, pp. 263–91.

—— (1997) 'The Invisible Hand and the Cunning of Reason', *Social Research* vol. 64 no. 2, pp. 181–98.

Vanberg, Viktor (1986) 'Spontaneous Market Order and Social Rules: A Critical Examination of F.A. Hayek's Theory of Cultural Evolution', *Economics and Philosophy* vol. 2, pp. 75–100.

Vaughan, Frederick (1972) *The Political Philosophy of Giambattista Vico: An Introduction to la Scienza Nuova*, The Hague: Martinus Nijhoff.

Vaughn, Karen I. (1984) 'The Constitution of Liberty from an Evolutionary Perspective', in J. Burton (ed.) *Hayek's Serfdom Revisited*, London: Institute of Economic Affairs, pp. 117–42.

—— (1987) 'Invisible Hand', in J. Eatwell, M. Milgate and P. Newman (eds) *The New Palgrave: A Dictionary of Economics* vol. 2, London: Macmillan, pp. 997–8.

Vernon, Richard (1979) 'Unintended Consequences', *Political Theory* vol. 7 no. 1, pp. 57–73.

Vickers, Douglas (1975) 'Adam Smith and the Status of the Theory of Money', in Andrew S. Skinner and Thomas Wilson (eds) *Essays on Adam Smith*, Oxford: Clarendon, pp. 482–503.

Vico, Giambattista (1999) *The New Science: Principles of the New Science Concerning the Common Nature of Nations*, 3rd edn, trans. David Marsh, Harmondsworth: Penguin Books.

Viner, Jacob (1958) *The Long View and the Short*, Chicago: Chicago University Press.

Walker, Graham (1986) *The Ethics of F.A. Hayek*, Lanham, MD: University Press of America.

Walton, Craig (1990) 'Hume's England as a Natural History of Morals', in Nicholas Capaldi and Donald W. Livingston (eds) *Liberty in Hume's History of England*, Dordrecht: Kluwer Academic Press, pp. 25–52.

West, E.G. (1964) 'Adam Smith's Two Views on the Division of Labour', *Economica* vol. 31, pp. 23–32.

—— (1975) 'Adam Smith and Alienation: Wealth Increases, Men Decay?', in Andrew S. Skinner and Thomas Wilson (eds) *Essays on Adam Smith*, Oxford: Clarendon, pp. 540–52.

Wightman, W.P.D. (1975) 'Adam Smith and the History of Ideas', in Andrew S. Skinner and Thomas Wilson (eds) *Essays on Adam Smith*, Oxford: Clarendon, pp. 44–67.

Wilhelm, Morris M. (1972) 'The Political Thought of Friedrich A. Hayek', *Political Studies* vol. 20 no. 2, pp. 169–84.

Wilson, James Q. (1997) *The Moral Sense*, New York: Free Press Paperbacks.

Winch, Donald (1988) 'Adam Smith and the Liberal Tradition', in Knud Haakonssen (ed.) *Traditions of Liberalism: Essays on John Locke, Adam Smith and John Stuart Mill*, St Leonards, NSW: Centre for Independent Studies.

Winch, Peter (1990) *The Idea of a Social Science and its Relation to Philosophy*, 2nd edn, London: Routledge and Kegan Paul.

Witt, Ulrich (1994) 'The Theory of Societal Evolution: Hayek's Unfinished Legacy', in J. Birner and R. van Zijp (eds) *Hayek, Co-ordination and Evolution: His Legacy in Philosophy, Politics, Economics and the History of Ideas*, London: Routledge, pp. 178–89.

Index

Lightning Source UK Ltd.
Milton Keynes UK
UKHW020136041222
413306UK00008B/45